A Soldier's Story

JOSEPH E. BAILEY

edited by Kaytlin Bailey

A Soldier's Story

ISBN: 978-0-578-27718-9

50 copies of this book were first printed in the USA by 48 Hour Books in December 2022. This second edition of the book, initially produced in January 2023 and then printed on demand, is in association with Kindle Direct Publishing.

Cover design and book layout by withleahmoon.com

TABLE OF CONTENTS

FOREWORD

Dear Reader,

My father was and always will be a constant source of inspiration and encouragement in my life. Every morning he used to make me breakfast and give me pep talks on the way to school. On days it was hard to wake me up, he would say, "You can't soar with the eagles if you hoot with the owls." I was a natural night owl.

My father did everything in his power to give me as many opportunities as he could. He also raised me to make, and live with, my own choices. He was honest about his mistakes. He shared some of his more painful lessons in an effort to save me some time. He respected those moments when I had to learn things the hard way, which was often. I know he is proud of me, and I am proud of him.

He has led an exceptional life. Undeterred by the limiting circumstances he was born into, my father achieved what he set out to do. Toward the end of his life, he said there was nothing on earth that he wanted to do that he hadn't already done. I believe him.

I became a storyteller because my father was a storyteller. I grew up watching him mesmerize people at parties. He told me stories about his childhood, invented tales for bedtime and long drives. Once, he was so engrossed in the story he was telling me that he drove the wrong way on 1-95 for over an hour. When he went to Desert Shield/Desert Storm, he recorded his stories on tapes for my mother and me. The ones for me were all beautifully told bedtime stories.

As I got older, he told me more complicated stories, about war and the women in his life.

After his retirement, I asked my father to write down his stories, to preserve them for the next generation, and as a gift to remember him. This is what he wrote for me. The book is in his voice, but the occasional footnotes are in mine. Of the many gifts he gave me, trusting me with his story is the inheritance I will cherish the most.

– Kaytlin Bailey

This Is My Life Story

I have not studied philosophy, psychiatry or the human condition. But I have lived from 1948 to now[1] and in that time I've been a participant in, and witness to, much of American history. Now in my early seventies, I am writing about my life for my daughter because I promised her that I would[2].

I am not a particularly religious man, nor do I claim any profound knowledge about what is right or wrong. There are many paths a person may take in life, each with their own risks and potential rewards.

These are the significant things I remember. I hope you have time to talk to your mother, who shared a large part of it. Your aunts, Ruth Ann[3] (AKA Renee), and Joanie,[4] and my adopted sister Sondra Woods can fill in any questions you may have. There is one other important person in my life, Rupert Finney, who lives in Ozark, Alabama. He has the distinction of being the only person with whom I have contact outside of my family and in-laws.

– Joseph Bailey

---

[1]My father died on August 31, 2022 from multiple myeloma, a blood cancer caused by exposure to Agent Orange in Vietnam. His sister-in-law, Deborah K. Mayer wrote his obituary.

[2]I gave my father his last mission, to write this book for me.

[3]Ruth Ann Dean, a.k.a. Renee, died on December 5, 2018, at the age of seventy-eight. She was born in Wichita in 1940. I stayed with her from November 29, 2018 until her death and wrote her obituary.

[4]Joanie Williams died November 26, 2016, at the age of seventy-two, just a few days after her husband. My father cared for them both in their home for three months until their deaths. My mother wrote her obituary.

Part I: My Childhood

Hutchinson, Kansas

My Father

I know the following about my father: he was born in 1889[5] in the Oklahoma Territory, where his family settled prior to statehood. He later joined the Navy and served in World War I.

When he signed up, he did so for the duration of the war. When his ship docked after it was all over, he packed up his gear and went home. After his death we received a letter from the Navy saying that he had left without being properly discharged and had been dropped from the ranks as a deserter. My mother told me he must have thought he had fulfilled his obligation and owed nothing more to the Navy. I guess they didn't see it that way.

My first memory of my father was when we still lived in the big yellow house in Hutchinson, Kansas. I was very young at the time. On this day, my mother and sisters were in the kitchen and the maids were in the living room. Butch, my dog, was not allowed in the house. I climbed the stairs and went into my parents' bedroom. My father slept most of the day and he was there—sleeping. I don't believe I had ever been in their room before.

He had hung his shoulder holster with a large pistol in it on the head of the bed. I managed to pull the whole thing off, hung it around my neck, and started down the stairs.

With every step the pistol hit the stairs and made a loud clunk. I was maybe three quarters of the way down when my father screamed "Ruth!" which scared me. I turned and there he stood, naked as a jaybird. His face was really red. My mother grabbed me in one arm and tossed the holster with the pistol in it up to him with the other. I do not remember anything else happening as a result of my invasion, but there was a big fence at the bottom of the stairs after that.

My father was a hard man, not given to displays of affection, verbal or physical. I cannot remember him ever touching my sisters, my mother, or me in anger,[6] but neither do I recall him ever hugging any of us children or saying he loved us.

---

[5]My grandfather was almost sixty when my father was born March 26, 1948.

[6]My father may not remember it, and his father may have never hit *him*—but everything I've heard from my aunts and Sondra, and what I know about violence in the family, suggests that my grandfather was a violent man both in and outside the home.

On Saturday mornings I often went with him to a pool hall on the second floor of a building on Hutchison's main street. The bar ran about half the length of the room, which was long and narrow, and the air smelled funny. Someone was always playing pool on one or two of the tables. My father would sit at a round table, drinking beer and talking with two or more old guys.

Meanwhile, I had the run of the place. My favorite thing to do was to take off my shoes and, in stocking feet, try to climb up onto one of the tables. This was not an easy thing to do. Dad always said, "No chairs or help." If I made it up to the tabletop, I would run the length of the table and jump. Someone caught me every time, but it was never my father, who was walking with a cane by then and moved too slowly to play.

Whoever caught me would swing me around and put me down and I would climb up all over again. By the time I was sent home to my mother, I was always exhausted and content.

One day I was playing at climbing the pool table when a couple of sailors[7] came in and told me to get off the table. My father told them to use another table. When they made a stink about it he beat them with his cane. Then he dragged them to the top of the stairs and threw them down so they rolled out onto the sidewalk. I don't remember trying to climb any pool tables after that.

My parents were bootleggers, buying whiskey out of state and selling it illegally for what must have been a significant profit.[8] My mother would get a call, deliver the whiskey and collect the cash. It was dangerous work. She was robbed and beaten by sailors on at least one occasion.

The next time I remember my father being a little testy was when he came to see us early one morning when it was still dark and we were all sleeping. My sisters and I all slept in the same bed at that point, and they must have woken me up.

He was sitting at the table and throwing a knife into the bathroom door, obviously angry about something.

My mother asked if we were hungry and we all said yes. My dad asked for scrambled eggs and bacon. He left the knife in the door when my mother told him to stop. He told us to sit down and then started

---

[7]During WWII, the US Navy constructed an Air Station thirteen miles outside of Hutchinson, KS, which was reopened for a few years in the 1950s before being closed in 1958. The base was then taken over by the Kansas National Guard.

[8]Kansas had prohibition from 1881 to 1948, longer than any other state, and continued to prohibit liquor in bars and restaurants until 1986.

telling us that the world was not easy and that we had to be prepared to take care of ourselves. People were not to be trusted. He went on and on.

As he was talking, he got the knife and started throwing it again. I was focused on the knife. The door was full of holes and it seemed a long way to throw a knife to stick it every time.

I don't remember what he said that made my mother use her "you are in trouble" voice. But I remember her telling him to watch his language. He said he was sorry and started imparting life lessons again. It wasn't long before he said something bad again. My mother said to "watch it" again, in the same voice, and he apologized.

The third time he cussed my mother didn't say anything—she just walked over and hit him in the head with one of her cast iron skillets with enough force that he fell backwards out of his chair and slid across the floor. He ended up with his head and part of his shoulders under the stove. From that position he softly said, "Ruth, I will not do it again. Can I get up and eat?"

I was sitting very still and watching my sisters to see if we should run or stay. They were sitting calmly, with no discernable expression of fear or worry.

A few minutes later, my father went into the bathroom and returned with a big wad of toilet paper on his forehead. He was still bleeding. My mom took a minute to clean his head, wrapped it with gauze, and served us all breakfast.

The change in my father was remarkable. Now he was smiling and seemed to be in great spirits. He ate his eggs with ketchup, Worcestershire sauce and Tabasco. He asked me if I wanted to try it. I said yes. He doctored my eggs and I have been eating them that way ever since, except when I am in public or a guest at other people's homes.

Every once in a while, he would come over in the middle of the goddamn night, drunk as a skunk, and beat on the back door. This was after we had come home from the orphanage, where my sisters and I were for about a year when my mother went to prison, so I would have been about four or five. After we came back from the orphanage my mother and father lived apart and we rarely saw him.

Another memory of my father—again, in the middle of the night. My mother came into our room and told us to get in the closet. Then she jerked the mattress off of our bed and stuck it on top of us. Then we heard loud explosions. Ruth Ann and Joanie told me to stay where I was and then they both crawled out of the closet to see what was happening.

There was a lot more shooting before I climbed out. The Sheriff (I found out later) called out and told my father to, "Come out, before anyone gets shot." My mother said something and pointed at my sisters and me. He looked at us, stood up and walked out the front door. When I woke up, Hiram Walker, a friend of my parents, was there with some other guys, fixing the windows and holes in the house.

I did not see my father for some time after that. The next time I saw him was, again, in the middle of the night.

My sisters were not in bed. I got up and walked into the kitchen where my father was bent over the sink. My mother was washing his head. When he stood up, I saw he had a small hatchet wedged in the side of his head. He told my mother to take it out and sew it up. Joanie and Ruth Ann were helping. Everyone was calm and acting like it was no big deal. So I did the same.

While my mother cleaned the wound, pouring antiseptic on it, and sewed it up, my father never whimpered or said much of anything. Afterward, he asked if he could get something to eat, so we all ate breakfast.[9]

My father had other quirks. Once, on a Sunday drive in the country, he stopped traffic in both directions to let a snake cross the road. My mother told us that the Indians had given him a deep feeling for certain animals and that was why he protected snakes when he could. He had lived in the Oklahoma Territories and had a relationship with Native Americans.[10]

On another occasion, we were going to the farm of one of my father's friends to pick up some corn and watermelons. I was riding in the back seat, a friend of my parents was driving, and my father was in the front passenger seat. He always carried a large semi-automatic in a shoulder holster, and suddenly my father pulled his pistol and told the driver to stop the car and back up, carefully. He slowly rolled down the window, reached out with the pistol, and fired. The sound was very loud. He got out of the car and returned with a decapitated pheasant, which he put on the floor in the backseat.

---

[9]Renee told me a story about my grandfather. He was playing poker with some friends at the house when all of a sudden, he got it into his head that one of the guys he was playing with owed him money. When he asked about it, the guy said he had already paid him. Everyone was drinking. Without warning my grandfather took his knife, or maybe it was a fork, and put it through the guy's hand sticking him to the table. My grandmother, exasperated, said, "Joe, go check your shoebox." After going and checking my grandfather returned, withdrew the knife or fork, and said, "It takes a real man to admit when he's wrong." No further apology was offered or asked for.

[10]There was a lot going on between the Tribes and white settlers in this part of the world in the 1890s.

He was always something of a shadowy figure, showing up out of nowhere in times of crisis and then suddenly gone.

One day I was at a public pool in Hutchinson with Ruth Ann and Joanie when my father appeared. I really don't remember how old I was, but I was small. He asked me if I could swim and I said no. He took me to a pool that was really for adults and told me to kick my feet and swim back to the edge. Then he threw me out into the pool. It was very deep. When I popped up, he was gone, so I started kicking my feet and all of a sudden, I was swimming.

When I went to find him and tell him, he was gone. Joanie and Ruth Ann never saw him. They thought I was kidding, but it is the truth.

Later I saw an old movie where John Wayne did the same thing with a boy who didn't know how to swim. Mom told me that is how they used to do it and assured me that he was watching from somewhere and wouldn't have let me drown. I like to think he was. But honestly, I am not sure.

The last time I saw my father was when I was nine or ten. We had left Hutchinson with our mother a few years earlier, but my father stayed. Ruth Ann, Joanie and I all decided to go see him. Ruth Ann drove. We hadn't seen him in years, but he was right where we expected him to be: standing on the sidewalk outside a bar, talking to some old guys. When we pulled up, we all said, "Hey, Dad!" and he said, rather testily, "How many times do I have to tell you kids not to interrupt me when I'm talking?"

Ruth Ann, Joanie, and I didn't have to discuss it. We all got in the car and drove away.

MY MOTHER

My mother was always a puzzle to me[11]. She was quiet, almost timid, and always courteous. She almost never raised her voice.

I don't remember her ever turning anyone away when they needed a place to sleep or something to eat, even when we were eating oatmeal three times a day. She often struggled to pay the rent, working from before I woke up till after I was in bed. She graduated from Business College in Wichita when she was a young woman, but was never able to get a job in business. She worked all her life as a cook.

My sisters were my primary caregivers and they were just kids themselves[12]. They would torment me, and fight with each other, unless

---

[11]Ruth Walker was born in Harper County Kansas in 1910. She married Joseph Bailey Sr in Oklahoma in 1944. She was thirty-four, he was fifty-five.

they got bored or I got in the way of something they wanted to do. Then they would hold me down and spread Vicks VapoRub under my eyes. This would force me to keep my eyes closed and I would eventually fall asleep.

They had other ways of immobilizing me too. Sometimes they would lock me in a closet or put me in a bathtub and take away all my clothes. A few times Mom came home to find me in the tub, my skin all wrinkled from being in there so long. It sounds like torture, but it was what they had to do to get some time for themselves and keep me out of trouble.

As a child, I was never really big on doing things with other people. I am still pretty much a loner. Trust is something I have very little of. I learned early that people do not really give a shit about anyone other than themselves. I hope this is not a surprise to you.

My mother was an outcast in her family. They all seemed to look down on her—and us. I do not remember her ever asking anything of them, or of them offering us any help. But I can remember wanting my mother to have a better life. In the end, she did.

My mother's maiden name was Walker. The Walker family comes from money. They all had farms, the whole clan. They moved to Kansas from, I want to say, somewhere in the Carolinas, Georgia, or maybe Ireland. They were wealthy before the Civil War, when they had slaves. There's a book that one of my great aunts wrote, *The History of Us* which gets into the history of the Walkers and the Kelloggs. My uncle Kellogg was about six-six. There's a picture of him, a big, tall, strapping guy. He was a farmer.

Kellogg was my mother's, mother's name. The family was wealthy, but my mother's mother married this guy named Walker who was building a railroad. He was one of the engineers, you know, not an engineer like we think of them today, but a guy who had a team of people that kept the rail line. His crew had given him a gold pocket watch one year, and he gave it to my mother Ruth, his oldest daughter, who was kind of his pet.

My mother had two sisters, Gracia and Dorothy, and they married well. They had a brother who died during the 1918 flu pandemic. Their father died in an accident on the railroad, I'm not sure when.

One of my uncles on my mother's side, a Kellogg, held onto the watch for a while because when my mother got married to my father, he was a bit of a rascal. Sometimes he'd come home and grab some of her jewelry, sometimes something he'd given her the day before. He would

---

[12]Renee was born in 1940, Joanie was born in 1944, my father, Joe, in 1948.

go out and pawn or hock it because he was losing at poker and wanted his money back.

So she gave the watch to this cousin to hold because she didn't want my father to take that watch. She didn't care about the other stuff, but her father's watch had sentimental value. After she left him when I was maybe eight or nine, we drove out to this place in the middle of nowhere. It was kind of crazy with pigs and chickens running everywhere. It was right out of a Tennessee Williams scene.

It was a big family with a lot of kids running around and this guy came out and my mother knew him right away. He was a cousin but I'd never seen him before in my life. He said, "This must be Joe." I didn't know what to say, you know? And she said, "I told you I'd be out here to pick up the watch." He said, "Well, it's around here somewhere." And my mother, who was a big woman, and strong, said, "I want the watch—now." This guy said, "Well, you know, I had to have it cleaned." She was looking down at him and said, "Go get the watch." So he left and when he came back he said, "Oh, here it is. But, you know, I paid to have it cleaned." She said, "Well, that's good of you."

The watch is a Santa Fe special. That was the railroad they worked on, the Santa Fe line. She gave me that watch when I got married the first time. And when I die, I'll pass it on to you, if you want it. I've never held it but a few times in my life. I've mostly kept it locked away.

***

My mother spent time with my sisters, teaching them how to act like ladies and set a table properly. She taught them to be gracious, courteous, respectful, and honest. (The last one did not take at all with Ruth Ann[13], but Joanie tried.) They were both able to be very ladylike when it was called for.

But all three of us inherited our father's temper. Ruth Ann, Joanie and I all have a point at which the demons come out, as you know firsthand. For those times, I am truly sorry.

---

[13]My father would sometimes refer to Renee as his "half-sister" but was vague about the details. I learned later that my grandmother got pregnant by her (married) employer when she was working as a domestic in Wichita when she was thirty. Whether she had an affair, was raped or seduced is unclear. She lost her job because of the pregnancy and shortly after giving birth to Ruth Ann (a.k.a. Renee), she started seeing Joseph Bailey Sr., who was twenty years her senior. Ruth Ann changed her name to Renee later in life, and I always knew her as Renee.

I had no contact with my mother and Ruth Ann for close to twenty years. It started when I was in Nam. They wrote letters to me that were not very positive and when Melodye (my first wife) and I went to visit after I came back from my first tour, they were rude. So I took my wife and left.

I do not remember how many times the Red Cross contacted my commander, who always ordered me to go and talk to them. They would say your mother or your sisters are trying to contact you. I would say, "Thanks, but tell them I am busy." One commander, while I was in Germany, gave me an order to contact them. I responded, "Sir, let's see if that holds up in my court martial." He had gotten three or four requests from the Red Cross and I think he was a little pissy.

Their letters telling me that Melodye was a pain when I was just trying to survive were not conducive to me maintaining my morale, or a positive attitude in the battlefield—which really was life and death. So after a while I just started throwing their letters away without reading them.

I am, I guess, a lot like my father, hard and unforgiving. It was your mother that finally made me reach out to them.

I stopped by my sister and mother's house in Orlando and explained that I was about to marry Donna and wanted us to bury the past. I said I was willing to put the past behind us, if they were. My mother and Ruth Ann said they were, and it was over.

As your grandmother aged, she became mentally withdrawn and eventually was placed in a nursing home with dementia. Ruth Ann saw to her care and visited her almost every day and at all times of the day and night. She would tip the staff, but if she found anything wrong, she would raise hell with both the management and the staff at all hours of the day or night.

I saw my mother maybe four or five times while she was there. I do not think she knew who I was, but she was still as sweet and gentle as ever.

Then came the call. She had been unconscious for almost ten days. One of us three—Ruth Ann, Joanie, or I—were with her twenty-four hours a day. On her deathbed, as she was passing, Ruth Ann, Joanie, her husband Dick, and I were standing at the foot of her bed. I think all of us were thinking of special moments in our lives when she had comforted us.

I know I was wondering if she knew how much she had given me. How she had been with me in my mind during some of the most frightening moments of my life, giving me strength, courage and the determination to see the job through no matter the consequences. And in that final moment, she opened her eyes and smiled at each of us. Then she was gone.

I think she was given clarity of mind in that moment so she would know she was loved and respected by all her children. Those final moments were a great gift for a remarkable woman and her children. At that moment I was a small boy again, and she comforted me.

I was momentarily transported to my youth and Mom was telling me, "You have the choice. You can become whatever you want to be. You just have to do it, that is all there is to it. Only you can make the choice to be whatever you decide to become."

I took her advice and did what I really wanted to. I joined the Army and became a Sky Soldier, served honorably in multiple wars and became a pilot. The adventures, the ups and downs, all met or exceeded my expectations.

MY DOG BUTCH

Butch arrived when I was very small and not yet walking. My father got him for me when I was probably about six or seven months old. I think he was a Basenji mix, his tail curled over his back, and he did not really bark. He kinda hollered. When I did start walking, he would let me hold onto him for balance. I would sometimes grab his ear or tail for stability, and he never let on that it may have hurt him.

When he first arrived, we lived in a big yellow house. The streets were dirt and we had a railroad track across the street out front, and a dirt alley in back. Somehow that dog knew I had to stay within the perimeter, I couldn't go to the alley, I couldn't go to the street and I couldn't go to the railroad track in front of the house. These were my boundaries and Butch would not let me cross them. If I approached the edge of the alley, or the railroad track, he would pull me down and get between me and my desired course of direction. If I was crawling, he would get in front of me and force me to turn. We were both kinda young at this point but he never hurt me.

When Butch was with me, none of the kids in the neighborhood, even the older kids, would mess with me because Butch defended me. The fur on the back of his spine would stand up and he had this really ugly growl. He just went with me everywhere. When my sisters were at school, or with their friends, Butch was with me. He would lay beside me when I played at building a fort or a racetrack for toy cars. I would explain what I was doing and he would look at me as if he understood every word. We were inseparable. I had a wagon and I would pull him up and down the sidewalk in front of our house.

He would also rat me out on occasion. If I climbed on the shed, or too high in a tree, or went into an old barn in our neighbor's yard, he would start howling and all my pleading and scolding were for naught. Soon enough a neighbor or older kid would come and pluck me from whatever I had gotten into. Of course, my mother would hear all about it and would tell me, again, not to do whatever it was I wasn't supposed to be doing.

My mother had a big barbecue built in the back, an outdoor pit, and she used to cook barbecued ribs and pork and stuff like that out there. She had this big butcher knife and this great big wooden thing that she cut meat on. And somehow, I got hold of her big butcher knife. My father and mother came out and saw me with it, but before she could take the knife away, my father said, "Watch this. Watch this." Because Butch was circling me, you know, nudging me, pushing my arm with the knife. And finally, I dropped it, and he grabbed that knife. It was a very good knife because we had money, then. My dad said, "This is the best dog I ever bought." But Butch took off with that butcher knife, which was one of my mother's favorites, and they never could find it.

Butch did not like my father.[14] When he came to our house Butch's fur would stand straight up and he and my father would glare at each other until my mother or one of my sisters put the dog outside.

I do not know where he went while my sisters and I were in the orphanage, but he was there when we came home. That was one hell of a reunion.

Butch gave his life for me the summer we got back. My sisters were in school and I was out back playing by the alley. I was building a city in the dirt with streets for my cars and trucks to drive on. Butch was lying beside me, watching and listening as I explained the layout, when a rock came whizzing by.

When I turned, a boy I had never seen before was throwing another one. I asked, "Why are you throwing rocks at me?" He said nothing, just picked up another rock, which hit me on my right ear.

When I started crying, Butch went after him. I called and called him but he was not deterred from his attack. The boy threw a stone at Butch, which hit him and he yelped. But Butch was on the boy in seconds, and he was not playing around. The boy screamed and started running to his house with Butch chasing after him, biting at his legs. Suddenly people

---

[14] I bet.

were running toward the boy and Butch stopped and walked back to me. I was crying and we ran to our house. I told my mother what had happened as she cleaned the blood off my ear and Butch's cut on his head.

The police came to get Butch. They brought some other guy who put a noose around Butch's head and dragged him to a truck. My mother was talking to the police officers. She seemed calm, but I was not. I was screaming and trying to tell them what had happened. She told me to go into the house. I was angry and scared.

When she came in, she held me in her arms, tears rolling down her face. I stopped crying. She looked me in the eyes and told me Butch was trying to protect me and that he loved me very much. I told her that I loved him too. Then she wiped her eyes and tried to explain that he had done something that was very brave, and also something that was not forgivable. That in protecting me he would have to pay with his life. He had bitten a little boy and that was not allowed under any circumstances. I was grief-stricken and inconsolable.

A knock came at our door and the neighbor next door came in. He had seen the incident unfold and had told the police what had happened. But there was nothing they could do about it. The police promised that they would make sure that their report noted what our neighbor had seen, but they explained that if a dog bites someone, the Humane Society requires the dog be put down. That's the law.

My mother offered to get me another dog but I declined. I thought about this off and on for a very long time.

I had a few dogs later in life, but none compared to Butch—not until your mother bought a worm-infested little lab pup, against my wishes, and named her Megan. Watching the way you and Megan became pals brought back memories of some of the best times of my life.

The Orphanage

When I was born, we were living high on the hog. My mother and father bought a brand-new car every year, big, really nice cars. We were rich, bootlegger rich. Until we weren't.

I was very young and no one explained to me what was happening. I don't think anyone ever did. We were in the house. Everything was fine. Then something happened when I was about four. My mom and dad left, or they were taken away. My aunt and uncle showed up. And we kids got in the car and they took us to the orphanage.

Neither my sisters nor I liked our aunts, uncles, or cousins—except for our Uncle Bob. He was always nice to us and seemed to enjoy having us visit. When he took us to the bus, he would give Ruth Ann some money and say, "This is for your mother," just like my father did. But then he would say, "Don't tell your aunt." Our cousins all acted snobbish toward us. I thought it was because they were better off.

The day we were taken to the orphanage was kind of strange. First, we were taken to our Uncle Bob and Aunt Dorothy's house, where we sat quietly, waiting for something. I was clueless, but I could tell neither of my sisters were in good humor. I had learned from experience to follow their lead, so I sat quietly beside them. They seemed to know something I didn't, which wasn't unusual. When asked if we would like something, my sisters said, "No thank you," in a curt manner. I stayed silent.

It was late in the day when our Uncle Cecil and Aunt Gracie arrived. They picked up our bags and we followed them to the car. We all got into the back seat. I sat in the middle between my sisters. No one said anything for a long time. My aunt asked if we would like a piece of gum and my sisters said, "No thank you," in perfect unison. I knew something was wrong, but I also knew not to ask.[15]

When we arrived at the orphanage, it was late and there were no lights visible in any direction other than on the very large building we were now walking into. It appeared to be in the middle of nowhere, and it was. When we went inside, I think we were greeted by three nuns. Our aunt and uncle went and talked with someone and came back shortly.

They asked, "Do you need anything?" My sisters answered, again as one, "No thank you, and goodbye." I went with one of the nuns and Ruth Ann and Joanie went with another. As we parted, they looked at me and both gave me a little smile, telling me it was going to be alright. I was not quite four.

The nun escorting me was younger than the others. I would guess she was in her mid-twenties and Spanish, I think. She asked if I had eaten and I said that I had not. She took me to a large dining room and told me she would be back. When she came back, she set a tray in front of me with a glass of water. I saw there were turnips on the tray and told her I could not eat turnips.

---

[15] I've gotten a lot of mixed stories over the years about what happened. Family legend is that both of my grandparents were arrested for bootlegging, and neither of my grandmother's sisters would take in my father and sisters. Another story is that my grandmother was trying to leave my grandfather, and had written some bad checks in the process. When she was arrested for the bad checks, my grandfather was unable, or unwilling, to take the children and so was the rest of the family.

She said, "Sure you can." I explained, "If I do, I will throw up." She said, "That's silly. Will you try them for me? Turnips are very good for you." I sat there and thought to myself that I had to do it.

I looked at the tray and decided to eat them first (the turnips) and then something else—quickly, so maybe I would not get sick. I ate them as quickly as I could and as soon as I had finished, she smiled and said, "See, I knew you would be fine." As she finished the sentence, I projectile-vomited all over the table and her. When I stopped, I said, "I am sorry, but I told you I can't eat turnips." She got someone to help clean up the mess.

She came back after cleaning herself up and gave me a peanut butter and jelly sandwich or a bowl of cereal, I am not sure, and a glass of milk. I kept telling her I was sorry, and she smiled and said, "That was not your fault. It was mine."

She took me upstairs and into a large room with maybe twenty beds on each side, most were empty. It was dimly lit. I could see that maybe nine or ten of the beds had occupants. She put me in the first bed you came to and helped me get in. It was very high. She told me where she slept and that if I needed anything to knock on her door. It had been a long day and I went right to sleep.

The next morning, I woke up to voices and saw that everyone was up and getting dressed, so I did the same. The nun I had met was helping some of the boys get dressed. When we were all dressed, she lined us up in the hallway. There were boys ranging from us very young ones to teenagers, each had a nun with them. We just stood there for a while. Then a whistle blew and we started walking toward the stairs. Once down the stairs, I saw the girls coming from the other direction.

When my sisters saw me, they smiled and I smiled back. We did not speak. We just smiled and kinda nodded our heads. We were back in the same dining room as last night. The little boys all sat at the same table.

The older guys lined up and served themselves from a cafeteria line. We were brought trays with a bowl of some kind of hot cereal and a glass of milk. I do not know what the cereal was. I've never had it before or since.

After we had eaten, we went back up the stairs in the same manner in which we had come, except in reverse order. Then we were told to wash our faces and brush our teeth. I followed the group. The bathroom had toilet stalls and sinks with stools on them for us little ones to use to reach the sinks. There were shower stalls at the end, I guess about ten. I copied the other boys and when we got back to our room, we made our beds. I followed their lead and waited by my bed.

When the sister came out of her room, we lined up in the hall and she led us to a kind of play area. I was not the least bit interested in playing with anyone. I missed my mother, sisters and Butch.

Every day was pretty much the same for a while. When the weather got warmer, I found out there was a playground with slides, teeter totters, and a metal dome-shaped climbing gym. I would climb to the top and just look around. We seemed to be in the country next to farms. I noticed there was a big barn with chickens, a fenced area with cows, and I think a couple of goats.

I never interacted with the other boys or girls when we went to the play area. The younger girls joined us, but I did not interact with them either. When I was sitting up on top of the monkey bars, none of the kids would come up there because I didn't take any shit. My sisters used to beat the hell out of me and I'd beat the hell out of them. My instinct from early on was, if you pissed me off, I would hit you. And these other kids weren't used to that, but sometimes they would come and try to get up on that monkey bar with me. Well, I'd climb down and go hit them or kick them out of here. Then this sister would come over and say, "Joe, you got to play with the other kids." I'd go, "Okay." But after a couple of times, the kids left me alone and so the monkey bars sort of became my hang out. When we'd go out on the playground, I'd climb up on the monkey bars, sit up there and look around, watch birds and stuff like that.

I spent most of the time trying to spot my sisters. I saw them a couple of times, either getting on or off a school bus. Joanie had her first communion there and I saw her do that. Her hair was shorter and she looked very serious, but she did smile at me when she finally saw me.

And I saw these ground bees, bumblebees on the ground. I saw one of them go in a hole and I thought "Oh, I'll trap him. " So I put my hand over the hole and it stung me. I didn't cry because my sisters told me not to cry. But man, it hurt like hell. My hand got all swollen. It puffed up and looked like just a big ball.

When we were going in for lunch, a sister noticed that my hand was swollen. She said, "What happened?" I said, "A bug bit me." So she took me to the dispensary and they said, "Yeah, it's a bee sting. At least he's not allergic or he'd be dead."

One day a plane came by. I mean, a little plane that was right above where I was, and no one else seemed to see it. I thought there were little people that flew little planes that only I could see. I really thought that. It was many years before I learned about model airplanes.

There were a couple of special days when we would eat outside, like a picnic. I can only remember two. One was with peanut butter and jelly sandwiches and a chocolate drink in a pop bottle. We could only have one, which we all drank immediately. One sister, I can't remember her name, seemed to spend a lot of time trying to get me to play with the other boys and girls and asking me questions. I simply would not answer any questions about my family.

I came down with chickenpox and was quarantined with curtains around my bed for a while. That was when my sisters snuck in to see me, late at night. They crawled in under the curtains to visit and it was great.

They asked me if I was okay and I told them I was fine. They told me that they were doing fine as well. That has been our way our whole life; we are always fine. Even if our world is crashing, we are fine. I think we get that from our mother. She never complained about anything.

One morning we were taken to the big barn and the man who took care of the animals gave us a tour and explained how he took care of them. He let those who wanted to pet the tamer ones.

A storm came in very quickly with thunder and lightning and it began to rain really hard. The cows were still in the pasture, so he ran out to bring them in. When he grabbed the metal gate to open it, lightning struck.

I was with the sisters and a few kids, standing just inside the barn. I saw the lightning bolt hit the gate and I saw the man go rigid and fall to the ground. I ran out to help him and the two sisters ran after me. As soon as I got to him, I was swooped up by Sister Michaels and carried back to the barn. The other sister took a cloth and covered him. Then we were all taken back to the main building. There was a funeral service for him that we all attended. There were a number of civilians[16] in the chapel that day as well. We never went back to the barn again while I was there.

I was never scolded for running out to the man, but I felt I had done something wrong.

A few days after the service, Sister Michaels was cutting my hair when she cut my ear and it started bleeding. I asked, "Did you do that because I tried to help that man who was killed?" She burst into tears and ran out of the room. Another sister came and started cleaning my ear. Sister Michaels came back and told the other sister she would handle it. She stopped the bleeding, poured iodine on it, and bandaged it up.

The iodine stung like the dickens, but I did not move or flinch. I thought, "I will not give her the pleasure of knowing she hurt me." I

[16] I'm confident he means people not directly related to the church or orphanage.

could not stop the tears, though. When she came around and saw my face, she started crying and picked me up in her arms and said she was sorry for hurting me, that it was not intentional, and she was not angry with me for trying to help the man. I said, "It's okay. I won't tell anyone." She started crying again. She walked me back to my room and put me in my bed and said I should try and rest. She was still crying. I was kinda scared but determined not to show it.

After that something changed with Sister Michaels. She would smile at me more often. It felt like she wanted to be friends. I remained standoffish; I did not want to be close to anyone while I was there. I just wanted to go home and be with my mother, sisters and Butch. I have no idea how long we were in the orphanage, but you can ask Joanie or Ruth Ann.[17]

We were there for one Christmas, which I remember was like a bad dream. They took us all and put us into the auditorium. The youngest kids sat closest to the stage. And they had this "magician," but even the four-year-olds could tell he was really bad. Nobody clapped, or anything like that. When he got done, there was a group of girls from the home who were dressed pretty nice and they got up on stage and sang some Christmas carols and someone played the piano. I think it was one of the nuns, and that was really nice. I mean, they were really good singers. At least I thought so at the time. Everybody got a little stocking with hard candy—and that was it.

When our mother came, I thought we were going back to the house. I remember she was driving a very nice car, but it wasn't hers. My Aunt Dot loaned my mother her car to come out and get us. The next day, mom took the car back. Now we didn't have a car and we couldn't go to the movies anymore. It didn't bother me but, boy, my sisters were down.

The house we moved into was much smaller than our previous house and was directly behind the church.

BACK TO HUTCHINSON

My sisters went to this parochial school right across the alley behind the church. I think the church owned the house, which was not really a nice house, because that's where Mom paid the rent when she had it. My mother became a cook, a short-order cook working in restaurants. And I remember she got a job at a hospital.

I went to kindergarten, and first grade. I think I was in first grade before we saw our Dad. He must have been in the slammer that whole

---

[17]They were there for just over a year.

time. I came home one day, and there he was. He was never a real talkative guy, unless he was scolding, or trying to teach you something.

The priest was Monsignor Reedy, an old guy with lots of quirks who stunk. My mother told me he had a colostomy bag after an operation to remove his intestines, and that is why he smelled bad.

This was a puzzle to me. How could a bag replace an intestine? And how can you live without your intestines? I wondered about this. One day I was playing in the alley and I heard these pops. I walked to the front of the church and there was Monsignor Reedy with a rifle, shooting at pigeons.

I was watching him when a policeman pulled up and told him to stop. The cop said something like, "What would God think about you shooting those pigeons?" Monsignor replied something like, "God does not want these birds shitting on his house." The policeman shook his head and drove off.

Monsignor noticed me and said, "What are you looking at?" I responded that I was curious about the noise. "Come here, son," he said, to me. Then he asked me my name and where I lived. I told him. He said, "I don't think I've seen you in church." I told him that I did not go to church. He asked me why and I told him, "My mother thinks I would be disruptive."

"Would you be disruptive?" he asked. I told him I did not know what 'disruptive' meant. He laughed and said, "I bet you would be at that." Then I asked, "How do you eat if you don't have an intestine?" He smiled at me and told me, "It is a challenge." I asked, "How does it work?" He told me and showed me the bag. I asked him, "Does it hurt?" He told me, "No, but it is inconvenient." I asked if I could help him shoot the pigeons. He said he was done for the day and that I should go home.

When my mother came home, I told her about our conversation. For some reason she was really upset. She told me that I was never to bother Monsignor again. If I saw him after that, I went the other way.

The house behind the church is the house where I thought Ruth Ann was going to kill Joanie. My mother had asked Ruth Ann to give the house a good cleaning. For us, at that time, a request from our mother was considered an honor. Ruth Ann, who was maybe ten or eleven then, made us breakfast and told us we were to stay outside in the yard while she cleaned the house.

Ruth Ann went straight to work, starting at the back of the house. She moved everything she could and cleaned every surface. Joanie got very bored playing with me and started stepping inside, to ask if she could help. Ruth Ann was having none of it and told her to stop bothering her and to stay outside. I knew enough to stay clear of them when they started up. Joanie sat on the front porch, and I could tell she was angry and thinking of a way to piss off Ruth Ann. I sat down by the tree in the front yard and just watched.

I don't really know how long she sulked, but Joanie was stewing for a long time. After Ruth Ann finished dusting, washing and making beds she started mopping, which was the last step. She started from the kitchen, where the back door was, and worked her way to the living room, to the front door.

As she was mopping the last bit, Joanie opened the back door and ran all through the house, with Ruth Ann chasing her and screaming at her. When they finally came out the front door Joanie was laughing hysterically. Ruth Ann's face was really red. She was screaming obscenities at Joanie.

They ran around in circles, out the front door and in through the back door. I don't know how many times, but I do remember wondering at some point why Joanie didn't run down the alley or anywhere else. At some point they went in the back door, but they did not come out the front. I waited. All I could hear was Joanie laughing. I crept onto the porch and looked through the door.

Ruth Ann had tied Joanie to a dining room chair. She was holding our father's hunting knife at Joanie's throat, saying, "I am going to kill you." Ruth Ann was as angry as I had ever seen her. It was scary.

I walked in and spoke softly, trying not to spook Ruth Ann. I told her that if Dad found out she had touched his knife she would be in trouble. No response. Joanie was absolutely nuts, laughing uncontrollably. I pointed out that the house still looked great and that nothing had been really messed up. I finally said, "If you cut her throat, you will make a big mess and Mom will be very disappointed." She finally looked up at me. Even Joanie stopped laughing. I didn't know until that moment that they felt the same way I did about our mother.

Ruth Ann's color faded from deep red. And Joanie told her she was sorry and that she just wanted to help. In a millisecond they were busy doing the floors again.

Their bond has always puzzled me. I know they cared about me but what they felt for each other is something very special.

They continued to tease me and we had some terrible fights. Sondra, Ruth Ann's oldest and closest friend, would, on occasion, participate in their torment. It drove me absolutely nuts and resulted in some real knock-down drag-outs but somehow, we never did any serious damage.

I did come close on one occasion. Joanie had done her usual teasing and I was mad and chasing her. She climbed up into the window above the kitchen sink. I got a dinner knife and threw it at her. The knife hit her in the forehead—blade first. Joanie, of course, started bleeding and screaming.

When my mother came home and heard about it, she sat me down and had the "girls are different from boys" talk. She explained, "You cannot hit girls or throw things at them because it can cause permanent damage. You can hold them down, or push them away to defend yourself but under no circumstances can you hit them with your fists, kick them, or throw things at them." I told her I was sorry and that I would not do it again.

I tried to keep my promise, but I must admit, sometimes I went astray. Both of my sisters took advantage of my handicap until I was around nine or ten years old. One day I came into the house and Sondra, Ruth Ann and Joanie, apparently bored, started telling me they were going to hold me down and tickle me. I said I was in no mood for their games.

They came at me, the three of them, but pretty quickly I threw Sondra on the couch then picked up Ruth Ann threw her in a chair and said something like, "This stops now or one of you is going to get seriously hurt." Sondra seemed really surprised. I do not remember which of them said it, but they all agreed that their teasing and pestering me was over. The verbal fights continued but the physical fighting stopped.

KINDERGARTEN & FIRST GRADE

We lived in the house behind the church for quite a while.

I started kindergarten at the public school, not the parochial (Catholic) school where my sisters went. I asked my mother why and she told me that Catholic school was not for me, I guess she was right.

I was so bored in school I thought I was in some kind of school for slow kids. The teacher wanted me to play with blocks, or color some drawings. It was only a half day and they had us lie down on these mats/rugs and take a nap mid-morning. I had to lie there and be quiet. Some of the kids actually went to sleep. I finally asked my mother if I was stupid. She told me, "No, why do you ask that?" I told her about the school and how dim-witted the kids and teachers were. I did not go back to kindergarten.

When I went to the first grade I fell in love with Ms. Spurlock. She got married somewhere in there and changed her name. Anyway, she was just as sweet as she could be to me. When she asked me to do things I tried to focus.

But first grade was not much better than kindergarten. We did have recess, and could play outside on the overhead bars and the merry-go-round. But the classroom was absolute torture. The book they had us learning to read from was about Dick, Jane and their dog Spot. I just could not get into it.

I started daydreaming about adventures and soon enough, it was time to go home. I was never really confronted about it. I know my mother was clueless. At the end of the year, it turned out I was failing. When my mother found out that I flunked the first grade she was really upset. I could not have cared less.

One day I got home from school and my dad was just there. My sisters did not seem pleased to see him. I didn't know why, they never talked to me about that stuff. We walked in and I said, "Hi, Dad," or something like that. He picked me up and gave me a hug. He also gave me some kind of toy wooden gun. And then he left the room for a while. I think he was crying. I never really saw my dad cry.

My parents never lived together again. I think my mom blamed him for us being taken away.[18] But after that, every Saturday, he would come and pick me up and take me to town.

He would also come around when mom was at work. He'd come in and sit down wanting to engage in conversation. My two sisters would give him the cold shoulder but I, at five years old, would have to sit there in the living room and listen to him. Jesus Christ this man could talk.

He'd come and pick me up every Saturday like clockwork. He'd take me down to a barbershop where I would get a crew cut while he talked to the barber about me. Then we'd walk and he would tell me life lessons. "Never do this. Make sure you wash your face every morning. Always keep your fingernails clean." Just on and on. My fingernails hurt like hell when he cleaned them, because he didn't dick around. If something was in there, he got it out. I never cried. But he really went after those nails. I learned to clean my fingernails before he picked me up.

Our father knew he didn't have much time with us because of his age. So he was trying to sneak in and give us all the lessons he'd learned in his life. I think during this period of time he brought us a record player and all these records about learning German. He would make us sit there

---

[18]I support his mother's decision. My Aunt Renee alluded to the fact that her stepfather, my father's father began sexually assaulting her when she was twelve, which is one of the reasons why she got married so young, and one of the reasons why my grandmother left him.

listening to them, trying to get us to learn German. He didn't bring us any music.

When school started, I had a different teacher for my second attempt at first grade. When recess came, I was doing my thing on the overhead bars when a few kids I was with the previous year came up to me and started calling me a dummy.

I was surprised by their lack of fighting skills. This was my first of many trips to the principal's office. Things were not going well and Mom was not happy.

But luck was with me: we suddenly moved to Wichita.[19]

[19]I don't know why the family suddenly moved. I know that Joseph Bailey Sr. did not come with them.

Wichita, Kansas

Second Grade

I don't know how I made it to the second grade, but I did. The second grade seemed like a repeat of the first. I was bored and I mentally escaped every day. That went on for a while until one day I heard my mother screaming my full name. First, middle and last. Joseph. Edward. Bailey. I had no idea what I had done but I knew it was bad. Then she called me again. This time there was a break in her voice, like she was hurt or something, so I ran home.

She was crying. She had a letter saying the school board was going to test me next week because they thought I was unable to learn and that I might need special education. She pleaded with me to please, please, try to concentrate when I took the test. I promised I would.

The day of the test, a lady walked into my classroom and told the teacher I was to go with her. She led me to an empty room and explained why she was there. I was not concerned, although I probably should have been. She verbally gave me a situation and asked me to tell her how I would handle it. Sometimes she wanted me to draw it out.

I remember in one situation, I was supposed to have played in a stadium all day and just realized I had lost the key to my house. How would I go about finding it? There were many of these and I had to take a pencil and paper and show and tell her how I would do whatever she had asked me to do. Then we did puzzle games with blocks and rings and so on.

She brought lunch for us both and we talked while we ate. She asked me if I liked school and how I got along with my mother and sisters. She asked what I liked to do, if I was more comfortable doing things alone or in a group. I remember that day being just really fun. When we finished school was already over and all the kids were gone.

I am really not sure how long it took to get the results. I had forgotten all about it when my mother screamed my name again. It sounded to me like I might have flunked the test, but when I walked in, I knew I had passed. Mom was sitting at the kitchen table, smiling. "They say you are extremely smart. Your poor grades are because you're bored." I thought, "How the heck did they find that out?'' I had never told anyone how boring school was. But my mother was happy and so everything was okay.

NEW MEXICO

I did not finish the second grade or go to the third. We were in bad shape financially and Ruth Ann was moving to New Mexico with her husband. Joanie and I were to go with them for a while. Ruth Ann was married when she was fifteen or sixteen.[20] James Chapman, who went by Jim, was a lot older than she was. He was in his late twenties, maybe thirty. I did not care for him much. He seemed oafish and dimwitted to me.

Their son Donald was born[21] maybe a year after their marriage. I think we are about five years apart.

So the five of us—myself, Joanie, Ruth Ann, baby Don and Jim—all got in Jim's car and drove to Farmington, New Mexico. Jim rented a small Airstream trailer. He was with us for about a week before he left to go to some job somewhere. We were left to fend for ourselves. I do not remember him coming back very often. He would send rent money, but not really enough to buy food, groceries or clothes.

We became bold thieves. Ruth Ann and Joanie would make the list of things we needed, including sizes. Ruth Ann and baby Don would wait outside in the car while Joanie and I would take the list, go into the store, pick up the things on the list and walk out to the car. Then we would drive away. We would go from one store to another and simply walk out with the items we wanted. I remember on one occasion we picked up an iron, but it was the wrong kind so we returned it on our next trip to the store.

At the time I really did not think about it. But in hindsight, I have no idea of how we got away with it.[22] We got groceries, clothes, sheets, towels—you name it, we stole it.

When fall came and school started, Joanie and Ruth Ann decided we would all take the year off, which was fine with me. We had lots of time, Joanie and I, to play and explore. We antagonized a bull, climbed to the top of an airfield. We rode our bikes everywhere and got into all kinds of scrapes. I loved it.

[20]She was fourteen.
[21]Donald Chapman was born in 1956, his brother Steve in 1958. Ruth Ann also gave birth to two stillborn twins during this time which was very traumatic.
[22]White privilege and guardian angels.

FOURTH GRADE

The following summer our mother wrote and said it was time for Joanie and me to come back to Wichita. Ruth Ann bought the bus tickets. I remember her and Joanie crying when we left.

When we got into Wichita, Mom picked us up in a pink convertible Cadillac or Oldsmobile. I thought we were rich! It turned out that the car was borrowed and we had to stay at a boarding house for a while. Joanie slept upstairs, and I slept with my mother.

We were only there a few days before we moved to an apartment above the restaurant where my mom cooked. Joanie and I ate most of our meals there. School started and Joanie enrolled in eighth grade and I enrolled in fourth.

My only problems were cursive and multiplication. We were supposed to be writing the alphabet in cursive, not printing it. I was stuck for a few minutes before I noticed that they were both printed on cards above the blackboard. So I copied what was there.

When I got home after the first day I asked Joanie about cursive. She said, "You should have learned that in the third grade but you skipped it." So she taught me how to write the alphabet in cursive, which worked. My other big problem was that I still could not read. In hindsight, I am amazed that I got through school without anyone saying anything.

By this time, however, I had bigger problems than school work.

It happened like this. We had come in from recess and were lined up to get a drink at the fountain. This older boy was not letting a smaller boy drink. I walked to the front of the line and said, "Let him get a drink. We are all waiting." He said something and shoved me.

I've always had a problem with bullies, I just won't tolerate it. I went at him with everything I had. We fought all the way down the hall and hit the fire escape door. It opened and the alarm went off. We fought our way about a quarter of the way down the stairs when I caught him with a perfect uppercut. He fell back, lost his balance and tumbled down the rest of the stairs. When he landed, he screamed and started crying.

A couple of teachers appeared. One went down to him and the other took me to the principal's office. I was sitting outside his office when a teacher I didn't know brought in the kid who'd been trying to get a drink at the fountain to talk to the principal. My worst fear was them calling my mother.

I have no idea how long I sat there, but it seemed like a very long time. The teacher brought the other boy out and he sat down across from me. He had obviously not been paddled but he had a serious look on his face, as I am sure did I.

When it was my turn to go in, the principal was looking down at some papers for a while. When he looked up at me, he said, "So Joe, what is your story?" I told him what happened. He said, "Joe, you are not supposed to resolve issues with your fists..." And on and on. I was getting tired listening to him talk when he asked, "Do you have anything else to say?"

I thought for a second and told him, "I will not be bullied nor will I tolerate others being bullied." Or words to that effect. He looked at me for what seemed a long time and said, "Go back to your class. In the future, call a teacher if you see any trouble." I was expecting big trouble, but it never came.

I suspect that the teacher and the boy had come to my defense. When I found out the bully had a broken arm, I worried about the police for almost a week, but nothing ever happened. I never told my sisters or Mom about any of this.

SIXTH GRADE

The sixth grade was my Waterloo. On the first day after Miss Breising took roll, she pulled the shades down and started showing us slides of her summer trips from all over the world. She had slides of Europe, Egypt, South and Central America, Australia, the Philippines and my favorite, Africa. Then she said, "I will give you all permission slips for your parents to sign and on Fridays, if your parents allow you to stay an extra hour, and you keep up with your school work I will take you around the world and tell you stories from my travels."

I was really excited. This was a dream come true. I got my mother to sign the slip and took it back. When Friday came I could not wait until class ended. The bell rang and a few kids got up and left.

She pulled the shades and then asked me to meet her in the hall. She shut the door and asked, "Do you remember what I said about viewing the slides?" I was perplexed because I got my mother to sign the form. She clarified, "The part about keeping up and participating in class. Do you think it's fair that the others have worked very hard to do their assignments to watch these slides, while you have mostly daydreamed through class and have not even tried to do any of your homework?"

She had me, fair is fair. I felt angry with myself for screwing this up. I really wanted to see and hear about her travels. I think she could see that I was really ashamed and hurt.

She said, "I will make you a deal." I was elated for a second. Then the hammer fell. "You will not be able to stay today. But if you really try next week, I will let you start next Friday. You understand that you will have to show me you are really trying?" I stuck out my hand and she shook it. Then she said, "It is up to you now."

I walked home slowly. When I got home, I asked Joanie if she would teach me to read. At first I don't think she understood. I explained. She was really kinda shocked. Then she asked me how I had made it to the sixth grade. I said they passed me, which they did. I didn't do very much, but they always passed me. Then she agreed to help.

All of a sudden both of my sisters and my mother were asking me to read this or that. I drove them crazy reading things out loud, street signs, store advertisements, anything and everything I could see. I actually won a couple of spelling bees in class and was asked to read in front of the class. I didn't want to do it, but I wanted to see those slides and hear her stories, which were fantastic.

It dawned on me years later that Miss Breising had played me. She was a wonderful and caring teacher.

TROUBLE

I liked to dance, and I often went to dances at the local church in the summer. One Saturday after a dance broke up, I saw this guy, Gee Gee, surrounded by three other guys. Gee Gee was a Mexican kid, maybe two or three years older than me, and he seemed like a nice guy. I walked over and asked him what was going on. He said, "They don't like me dancing with the girls." I asked, "Can I help?" The other boys started telling me to leave or I would get the same. It went down pretty fast. The three boys were not really all that tough and Gee Gee and I were both scared and angry. We meant to hurt them and we did.

They took off and Gee Gee said, "Thanks."

I was arrested the next day and had to go to court. The boys were from my neighborhood, so they gave me up, but they didn't know Gee Gee's name or how to identify him. The cops asked me who he was, and I told them I didn't know. When I went to court, I was convicted of assault and given probation and a membership to the YMCA.

Gee Gee and I never really became close friends, but we would say hello when we saw each other. About a year later, he saved me from a beating.

I was walking home from a movie when I was confronted by five or six guys. I had taken a shortcut through the Mexican neighborhood. I didn't realize that as a white boy that I had crossed a line and they thought they were going to teach me a lesson.

I always carried a heavy dog chain about nine inches long with a gear shift handle taped to one end. I pulled it out and let it hang at my side. Two of the guys had switchblades. I thought, "This is going to hurt." Suddenly Gee Gee came running out of one of the houses, shouting in Spanish. The boys backed off and put their knives away. I hollered, "Gee Gee, I'm glad to see you!" He smiled and said, "I never thanked you for not squealing on me." We shook hands.

I asked why they were angry with me and he said, "You're white." I didn't understand. "I know," he said, "But they don't like white kids. Take another route home next time. I might not be there." I told him thanks again and walked home. We would see each other from time to time. If he was alone, we would talk. If he was with friends, we would just nod.

Growing up, I had Mexican, Black, Japanese and even German friends. The last two were really not liked. It was a WWII thing, I guess.

Around this time, I was arrested for auto theft, or joyriding, as it was referred to back then. It happened like this. I was walking home when these guys I knew, Chuck and Robert, pulled up in a car and asked if I wanted to take a ride. I was glad to and got in the back. We drove out to the Boeing aircraft factory and back into town. We were riding down Main Street when a cop from our neighborhood, Officer Copeland, pulled us over.

I didn't even know they had stolen the car. I assumed, since Chuck was driving, that it was his dad's or a friend's car. When I went to court, both Chuck and Robert said it was my idea. And Officer Copeland told the judge that although I was younger than either of the other boys, I was the ringleader and was most likely responsible. No one asked me what happened. I was found guilty and given a stern lecture and another membership to the YMCA.

That made three times I was given a suspended sentence. (The first time was a year or so prior to the assault, when I stole a bike.)

The sentence made my mother cry, which really pissed me off. It took me about a week to catch up with Chuck, and then Robert. I warned both of them that if they reported the beating, I would hunt them down and kill them. I guess they believed me because I didn't hear anything about it again.

After that I was a loner. I still stole bottles to cash them in and prowled the neighborhood at night, but I was very careful not to get caught.

SEVENTH GRADE

I started the seventh grade with a bit of a bang. On the first day of class I recognized two girls I had known from years ago, twin sisters who lived down the street from me. I nodded at them and found a seat.

I was sitting quietly when a boy I had never seen before sat down next to the sisters and started talking with them. Other kids were likewise engaged in whispered conversation, and it all seemed to be about me. I just wanted to be somewhere else.

The boy the twins were talking to got up and said something like, “You think you're a tough guy, right?” I told him, “No, I don’t.” He called me some names and I got angry. The sisters said, “Look out, he’s getting mad!”

When I looked over at them the kid knocked me out of the desk. During the fight, he bit me, which sent me into a rage. I was sitting on his chest and hitting him in the face as hard and as fast as I could when the teacher walked in.

I stopped and stood up. The kid I’d been fighting with pulled out a pencil and I blocked his thrust with my hand. The point stuck in my finger. I do not remember what I said to him, but it included words that were inappropriate for a classroom. As he tried to get up, I grabbed his head and jammed my knee in his face. It was over when his nose started bleeding and he was crying.

The teacher grabbed me by the neck and sat me down. I tried to explain that I had not started the fight but he was not interested. He took us both to the boys’ room and told me to wait in the hall while he took the boy in. They came out after a while. The boy was holding hand towels to his nose. His eyes were red and swollen and his lip was cut. I couldn’t help smiling at him.

We were both taken to the principal's office.

The boy went in first. When it was my turn, the principal asked me why I had attacked the other boy. I explained that I had not attacked him, he had attacked me. I was sent out to wait outside the office with the other boy. We both watched as one by one, the other kids went in and came out. He and I sat there looking at each other. His mom came and got him and the look she gave me was not friendly.

Eventually I was called back into the principal's office. He told me the class had confirmed my story, however, that was no excuse for my vicious attack. I started to explain but he cut me off and told me to go to my next class. When I walked in, everyone, including the teacher, was looking at me. I told her I was sorry for being late and headed for an empty desk. When I sat down all the kids were still staring at me. I focused my eyes on the teacher and tried to ignore them.

The next day the teacher acted as if nothing had transpired the day before. The boy was not there. Later I saw him in the hall when we walked to our second period. He really looked bad.

Denver, Colorado

A month or so later my mother, Joanie, Ruth Ann, her two boys and I all moved to Denver, Colorado, to be near Sondra and her brother. We lived with them for a few weeks, until we got an apartment five blocks from Belmont Jr. High School.

My first day I met with the assistant principal, who was about my mother's age. I sat in her office while she read my school record.

I hoped she had not gotten the police reports from Wichita.

Apparently, she hadn't, because after she finished reading, she smiled at me and said, "I hope you will find Belmont to be a friendly school and take advantage of the opportunities we have here." I said I would try. Then she said, "Here we like our boys to have short haircuts." I said, "I like mine long." She stopped smiling and said, "I'll take you to your next class and introduce you."

I could tell it was not going to be a good day. The kids all seemed to stare at me. I was the only one wearing blue jeans and with long hair. This was 1960, and only rebels had long hair.

At lunch a guy named Brian introduced himself and volunteered to show me the ropes. As we walked out, a boy walked up to me and said, "I chose you." As he walked away, I asked, "For what?" But he just kept walking. Brian said, "That means he wants to fight you."

I caught up to the guy at the top of the stairs and beat him until he started crying. Then I stopped. I asked if we were through, and he said yes. Brian explained that I was supposed to meet him *after* school to fight. I felt bad for the misunderstanding. When I saw him in gym class later, I apologized and offered to meet him after school, but he said, "No thank you."

The next couple of days I met two or three other kids who wanted to fight on my walk home. I was always amazed by their lack of aggression. They always took up a boxer's stance but then just waited for me to throw the first punch. One or two and it was over.

I did have one run-in that I never figured out. One Saturday, Brian and I were walking down Colfax, which was a main road, when a car stopped. A guy jumped out, sixteen or seventeen years old, and he pushed me down. "Leave my girlfriend alone," he said. As I scrambled to my feet, he got back in the car and drove away. I never saw him again and never had a clue what he was talking about.

I had one other close run on my way to shop class. Two ninth-grade boys caught me outside, alone, and told me they were going to beat my ass. I dropped my notebook, took two sharpened pencils, one in each hand, and started walking toward them. They turned and ran away.

I just didn't fit in Denver. I had long hair and wore jeans and casual shirts. Brian said I should just get a haircut. I said that I would think about it, but I never cut it short until I enlisted about three years later.

Gym Class

After a while I stopped being a curiosity. Gym class was fun. We did exercises, climbed ropes and played dodgeball.

One day the coach told us that there was going to be a competition for sit-ups, pushups, and the rope climb the following week. If any of us wanted to compete we should let him know.

I had been doing sit-ups from a very young age while my mother or sisters held my legs in their laps. When something was bothering me, I would do them until I was absolutely exhausted. It took my mind off of whatever else was going on.

I didn't want to draw attention to myself but I thought it would be fun to compete. I asked if I could do sit-ups and the coach said what the school record was, and that I would have ten minutes to do as many as I could. Seemed straightforward to me.

In my mind all of this would be happening in our gym class, like a normal day. But the day of the event we went down to the locker room and it was full of students from seventh to the ninth grade. When we went back up to the gym it was full of kids and teachers. I hated being in front of so many people.

They started with pushups. And they competed one at a time. Then the rope climbs, again one at a time. I was trying to think of a way out when it suddenly dawned on me that I was the only seventh grader in any of the competitions. They went by grade senior to junior, which meant I would be the last competitor. I became catatonic. I felt like a total fool.

When it was time for the sit-ups, they put a mat down and the coach called the next guy up. I watched but really could not pay any attention. Finally, they called my name. I walked slowly so I wouldn't trip or something. I got down on my back and the coach held my feet and said, "You need to breathe." The girls' coach stood above me, and the wrestling coach had the stopwatch. Someone asked "Are you ready?" I nodded and the lady coach said go. I remember nothing until she said stop.

When it was over there were cheers and applause and my coach was grinning. The coach grabbed me, raised my arm in the air and said I was the new school record holder.

I was hot and flushed and just wanted to get out of there. My head was down and I was completely embarrassed. When they let me leave, I ran down the stairs without showering.

I did not go to school for two days. On the second night, my mother came in and said, "The coach called to see if you were okay." I said that I was fine, I just didn't want to go back to school. She told me I had to go, so I went the next day. The assistant principal sent for me in my first class. I had no idea what I had done.

In her office she asked me if something was bothering me. I do not know why, but I told her that I had thought it would just be our gym class and that I did not like being made to stand up in front of the whole school like that. I just wanted to try and do a lot of sit-ups, that's all.

She said, "Joe, you are a puzzle to me. You say you do not like attention, but you fight and keep your hair long and do not really try to get along with other kids. I do not understand. Can you explain it to me?"

I told her, "I do not bother anyone and expect to be treated the same. But for reasons I do not understand, kids want to make me do this or that. The only person in the world that I obey is my mother and I really do not care what anyone else thinks about me or wants me to do. I just want to be left alone. And if some kid wants to fight—I will not be bullied by anyone."

"I called you in to make sure you hadn't hurt yourself. You were very flushed and looked sick."

"I was embarrassed by all the cheering and stuff. I don't like it."

"I just want to tell you, Joe, I am proud of you. You can wear your hair any way you want and if you ever need my help, I'm here. You can go back to class now."

Nobody bothered me for the rest of that day.

CHRISTINE

Every once and awhile they had a dance at school. The first one I went to was great. All the girls would sit on one side in chairs against the wall and us boys on the other, in the bleachers. Someone would announce the next song and the boys would ask a girl to dance.

I would wait until all the guys had asked a girl to dance. Then I would pick one I thought no one would ask. I thought this was the proper thing to do. I never danced with the same girl twice and I danced every dance, including the twist and slow dances, which was unusual. The dance stopped at ten.

The last dance was a ladies' choice. I was getting my jacket and this girl tapped me on my shoulder and asked me to dance. As we danced, she asked me all kinds of questions. Like why I had long hair, why I liked to fight, and why had I danced with a different girl every dance.

I answered them. I liked my hair the way it was because my dad had made me get crew cuts every week until we left Hutchinson. I did not like to fight but I was not a guy who ran from one either. I told her that all the fights I had were started by kids I didn't even know. And I thought it was the polite thing to do since I did not come with a girl.

She said most boys would not dance with fat or ugly girls. I told her that was their problem not mine.

I really could not see her face. She had long brownish blonde hair and she kept her head bent, kinda like she was looking down all the time, or was very shy. The dance ended and I escorted her back to her seat and, as with all my dance partners, I thanked her for the dance.

Brian ran up to me as we were leaving. "How did you meet Christine?" I had no idea who he was talking about. "Christine Coronette," he said. I still did not know her. "The last girl you danced with. She is an eighth grader and the prettiest girl in school." I told him, "I don't know her. She just asked me to dance."

At lunch on Monday, Brian was on it again. "Did you see Christine today?" I was getting a little tired of it. "Look," I told him. "I am poor and different. She was probably just curious about the weird kid in school. Stop trying to fix me up." He finally stopped and things went back to normal.

There was a creamery about a block from the school that I would go to with Brian sometimes to get shakes or ice cream. One day we walked in, and Brian said, "There she is." "Who?" I asked. "Christine." I said, "Will you get off that?" We ordered our shakes and were leaving when she said, "Aren't you going to say hello?"

We turned and Brian said hello. I said, "Do I know you?" She said, "We had a dance." "Oh, I'm sorry," I said. "Hello." And then we left. She came out and said, "You know you are being rude." I said, "I was not trying to be. What am I supposed to say?" "You could come in and sit with us." I looked at Brian and he said it was okay with him, so we sat down with her and another girl named Nancy. Christine's hair was still hanging over half her face but I could see that she was very pretty.

I responded to the questions I was asked. Brian did most of the talking. When we finished our shakes, I was more than ready to go but Brian just kept on talking. Finally the girls said that they had to go. We all got up and walked out together. Brian and I lived in separate directions, the girls were going his way. I said, "see you tomorrow," and went home.

The next day, Brian was very excited to tell me that Christine liked me. I was not at all interested. He said, "She is smart, beautiful and she likes you. What's the problem?" I explained once again that I was in no position to ask anyone for a date. "Brian, you should ask her out. You always have money. You could take her to a movie or something. I am not interested."

Note: I have never been much of a ladies' man. They have always puzzled me. I have never been really comfortable with females. I listened to my sisters talk about their boyfriends and guys about their girlfriends and it seems to me that females really do not like males. It is like we are a different species. I have never approached any girl or woman, including your mother. I notice attractive women, but it is not in my nature to try and get to know anyone, especially women. We do not really have all that much in common.

A few weeks went by. I sometimes saw Christine in the hall and nodded to her and Nancy. They were really the only two girls I knew by name.

At the next dance I told Brian he should dance with girls who had not been asked. So we would wait and then go and ask them to dance. We were having fun. When the third or so dance was announced, we were walking to ask someone to dance when Christine walked up and asked me to dance, which was not considered ladylike.

She had broken with tradition, and I was momentarily dumbfounded. I asked her if she was supposed to ask before the last dance? She said, "No, but it was obvious you were not going to ask me." I said, "Listen, I do not come from a wealthy family. In fact, we are dirt poor." She looked at me and said, "So. Are you going to dance with me or not?" I suddenly realized we were standing in the middle of the dance floor not dancing.

So we danced. It was a slow dance and she kept asking questions and I kept answering them. We danced every dance.

Afterward, I walked outside and took a deep breath. As I was exhaling, she came up and said, "Walk me home." So I did. We talked all the way to her house and stood outside talking until her mother came to the door and said, "Christine it is midnight." I apologized to her and her mother, and then I went home.

That was the first time in my life I ever talked with anyone that long. Or about my dreams. We became friends and talked on the phone. She would listen to me and I to her. Her family moved and I lost contact with her for about four or five months. When she called me, it seemed we had not missed a beat. We talked on the phone for about a year. She called to invite me to her birthday party and then said that I could bring a date. That was a blow because I had no date.

I sometimes babysat for a woman, Beverly, who lived in the same apartment as my mother and I. She found out from my sisters that I did not want to go to Christine's birthday party stag, so she volunteered to be my date. It seemed like a good idea at the time.

Beverly was a knockout. When we got there, we were shown to the basement. But it was obvious that we were overdressed and way too mature for the crowd. This was Christine's fifteenth or sixteenth birthday. All the boys had their mouths open when we came down the stairs. Even Christine looked wowed. I said, "We can't stay. I just wanted to drop off my gift. We have a dinner date." I gave Christine a hug and we left.

When we were driving away, we both started laughing. I saw Christine once more prior to leaving for basic training. We spoke for just a few minutes. I think she was my first love.

AL GREGORY

We moved to some other part of Denver that summer and I started the eighth grade in a new school. I went for a while, maybe a few weeks. But I told my mother pretty quickly that my school days were over.[23]

My mother had started asking me what I wanted to do with my life when I was maybe five or six. Ninety-nine percent of the time I would say that I had no idea. She would say, "You better start thinking about it because you're not getting any younger." Then she would smile and tell me, "You know, it really is all up to you. You can be anything you want to be if you just make up your mind to do it. That's all there is to it."

We were what most people would refer to as poor white trash. We lived day to day, relying on each other to get through the dark times—which we had more than our share of. She was absolutely correct: I needed a plan.

When I was fourteen or fifteen I got a bunch of minor jobs: McDonald's, a Pepsi bottling plant, laying carpet, etc. At one point I moved into an apartment with an older guy, who was about nineteen years old, for the summer. We worked our asses off just to make rent and eat. I started thinking hard about the future and myself. I found myself to be lazy, vindictive, self-centered, untrusting, unforgiving, hot-tempered, friendless, and scariest of all, planless.

---

[23]By this point both Renee and Joanie were married and out of the house.

Denver is where I met Alex Gregory, who was in his early thirties. He's the guy who painted your Aunt Ruth Ann and her sons' portraits. Al and I used to get a couple of six-packs and talk about life. We had many all-night talks. I would listen and ask questions as he gave me his philosophy of life, love, and death. I think he was the only person other than my mother who said, "You have to make your life what you want it to be; no one else knows what you want. It's all up to you."

My mother got a new job as a cook at a nursing home and got me a job washing dishes at the same place. I moved back in with her and gave her my paycheck. She needed the money more than me, I was happy with a few bucks for gas and beer so Alex and I could cruise around.

It was on one of these nights, driving around Denver with Alex, that I met Jeanee.

We were on Colfax around two a.m. when a lady stumbled out of a bar and walked right in front of the car. I stopped and she fell down in the street. I jumped out and asked if I could drive her home. She seemed really drunk, but she said she would like a ride. I helped her into the back seat and asked her where to go.

She was staying at the Red Wing Motel by the airport. I woke her up when we pulled in and helped her up the stairs to her door. She unlocked it and when I said good night she said she was a little afraid and could I please just come in and make sure everything was okay? I said sure, unlocked the door, walked through the room and the bathroom, and said, "I checked, everything is okay."

She said, "Would you just come in for a minute and make sure?" I said okay and walked through the room again, and started for the door. She finally asked if I could just stay the night because she was afraid. I thought about it and said, "Give me a minute." I went and told Al that something had her scared. "I said I would stay and watch the door until the sun comes up so you can take the car and pick me up tomorrow." Alex said, "I hope you know what you are doing, see you in the morning."

I went back and knocked softly, hoping she would be asleep. She unlocked the door and said, "Thank you." I moved a chair in front of the door and sat down. I said, "I'll be here until the morning, so you can relax and get some sleep."

That's when she asked me if I was still a virgin. I was shocked but I said yes. Then she blurted out, "Would you like to not be a virgin in the morning?'' I was absolutely dumbfounded. I just sat there looking at her. She was in her mid-twenties, tall, thin, with short black hair and very attractive. I told her, "I have never been with a girl and you are a woman.

I don't know what to do." She smiled and said, "I'll teach you, how's that?" I know this sounds like bullshit, but it is absolutely true.

She was very tender and explained everything she wanted me to do and what she would do for me. We made love all night in the bed, the shower, the floor and a chair. She told me what women liked, and what they did not like. She told me slow and tender was always better than fast and hard. I asked questions, which she answered, and then demonstrated, all night. We finally fell asleep as the sun was coming up.

When I woke up, she had gotten us coffee and rolls for breakfast. I ate quickly because I was very hungry. She asked if I would like to come to California with her. That's when I remembered that I had lied to her. I had to tell her how old I was.

"I'm only fourteen, not eighteen." Her eyes widened a bit. "You are only fourteen years old?" She did not seem angry but very unsettled. I told her yes. Her face got very red. She lowered her voice and told me, "If you tell anyone about this I will go to prison." I asked why and she explained.

I promised her I would never tell. I kept my promise for almost fifty-three years. I have thought of her fondly, often, for many, many years. As for her instructions, I took them to heart.

After I joined the Army, Al and I kept in touch. I think it was about a year or two later when he mailed me a *Playboy* magazine with a note to look at a photo layout.[24] There she was, as lovely as I remembered her.

DRIVING WITHOUT A LICENSE

I had started taking my sister's car and driving around town when I was twelve. When I was stopped by the police, which was pretty often, I would reach for my wallet (which I did not have) and say, "I'm sorry, sir, I must have left it at home."

The officer would ask me how old I was and I would lie and say I was eighteen. The cop would then cite me for "driving without a license" and I would throw away the ticket. I really have no idea how many of those I accumulated. My guess would be twenty-five or thirty by the time I was fifteen. I was never stopped for a driving infraction.

---

[24] I am still looking for this magazine, I haven't found it but I believe him. Growing up my father maintained that "girls obviously mature faster than boys" and also that he was ready for this experience. Of course, had he known about anyone coming on to me at that age he would have shot them on sight.

One day when I was about fifteen, I was working at the nursing home and my mother had the day off. Two men came in and asked me if I was Joseph Bailey. I said I was and that's when they arrested me.

As I was being booked, a guy sitting beside me asked if it was my first time. I told him it was. He told me to hide any money I had in my socks or shoes so that the guard could buy us soda or cigarettes for later.

I was handcuffed, taken to county jail, processed and put in a cell with two other guys in their twenties. They were playing poker and I asked if I could play. They said, "Sure, if you have any cigarettes." I did not, as I did not smoke. "If you have any money the guard will get some for you."

I asked the guard and he said, "Do you have any money being held for you?" I told him "No, but I have some in my sock." He got a little upset, but he took my money and left. When he came back, he gave me three packs of cigarettes.

We started playing poker and they asked me, "What are you in for?" I said traffic tickets. They started laughing. I said, "What are you in for?" It turns out I was in the slammer with the Purple Phantom[25] and a cop killer.

The Purple Phantom had been raping women waiting for buses for a few months and the other guy had shot and killed a Denver police officer a few days earlier. We were in a cell with about twenty beds and it was just the three of us. We played until lights out. (They never really turned the lights out, just dimmed them.)

The following morning, we got on a bus that had about twenty guys on it. I asked where we were going and they told me we were going to the maximum-security cell. I thought, "Wow, I'm in really deep shit." We had chains around our waists with handcuffs and ankle cuffs.

When we got to the county prison, I was in a cell with about sixty to eighty men. The two guys I was with the previous night went somewhere else. The next morning my name was called and I was put back in cuffs. When I asked the guard, he told me I was going to court.

It was a long drive and I had not slept the night before, so I dozed until we got there. When we arrived, I was put in a holding cell with one other man but there was room for around ten. We were both shackled to a large steel anchor in the center of the room, which was round and the walls were stainless steel.

---

[25]I tried to look this up but didn't find anything immediately. This was in Denver, CO in 1962 or '63.

My Mom was probably going nuts. I had not asked to make a call because I had no idea what I would say.

I sat there for about two hours, talking with my cell companion, who seemed like a nice guy. I asked what he was in for and he answered casually, "Murder." "Tough break," I said. We talked until a cop came and took me to the courtroom.

When we walked into court, it was already in session. There was a woman judge and she glared at me. Almost everyone was looking at me in shackles and my orange suit. The judge seemed to be really pissed. She concluded the current trial and called for a two-hour recess.

A few minutes later a man came in and told the cop the judge wanted to see him in her chambers. He left me in the hall with a warning about trying to run. I could only take baby steps and I would look really out of place running down the halls with no idea where to go. So I leaned against the wall and waited.

You could hear the conversation from inside. It turns out that the judge was not pissed at me, but at the cop for bringing me into her court in shackles. She dressed him down for a few minutes then told him to send me in, minus the hardware. She said she would send for him when she was finished talking to me.

I went in and she asked me to take a seat. She smiled and asked me how I had ended up here. So I told her the long story of being arrested and jailed with the Purple Phantom and the cop killer, then going to the county prison. I told her about taking Ruth Ann's car and getting stopped for the last three or four years and that the reason I had no license was that I was not old enough to get one. I wasn't sixteen yet.

This last part seemed of interest to her. "When were you born?" she asked me. I told her. Then I asked if it would be possible for someone to call my mother and tell her I was okay. She asked for the number and my mother's name, and if I would like a Coke or something. I said I would. She called the cop in and told him to go and get me a Coke. Both he and I were dazed by that. Then the judge left me alone in her office.

A few minutes later the police officer came back with a Coke in one hand and shackles in the other. He took a position behind me. When she came back, she asked him why he was in her chambers and told him to wait in the hall as she had asked him to do. This guy was going to beat the hell outta me when this was over.

She said, "I will let your mother know you are doing okay." Then she called him in and told him to take me back to holding.

When I got to the holding cell, I was the only one there. The cop said, "Why are you back here?" I said, "The judge sent me back." He said, "I know that. Why did she send you back?" I told him. "I don't know."

He said, "I have to sit here if anyone is in here. And I have other responsibilities, so how long are you going to be here?" I told him I was sorry, but I had no idea what was happening next but whatever it was wouldn't be my decision.

After some time, another officer came to get me. He brought my clothes in a large plastic bag. He took me to a locker room and told me to change into civvies.

I came out and he took the bag with the prison garb and escorted me back to the judge's office and knocked on the door. My mother was there, crying. I gave her a hug and tried to tell her why I was in trouble. She said, "The judge told me." I said, "I'm sorry. I had no idea what I was doing was such a big deal. I just liked driving the car."

The judge asked me to sit and told us she was going to let me go, but that I had to make monthly payments. She was also going to void the interest for not paying the fines on time. "When the fine is paid, your record will be expunged," she said. "But failure to make the payments will result in a reversal and you will be liable for the full amount."

I agreed to the terms. On pay day I went to the courthouse to pay. I told the teller my name, and the reason I was there. He knew nothing about it. I gave him the judge's name and explained the seriousness of my situation. He said, "Wait a minute," and was gone for some time. He brought back some man in a suit and I told my story again. He said he'd called the judge and she was not available, but he offered to write a memo and took my name and address and promised that if I owed anything he would send me a letter and explain to the judge what had happened.

A few days later I was out with Al Gregory and told him what had happened. He started laughing and said, "You've been had. They broke so many laws arresting you as an adult and putting you in the county jail that you could sue the shit out of all of them." I said, "If they expunge my arrest record, I'll be happy."

They must have, because later I got a NATO top-secret clearance and that incident was not on the sheet. All my juvenile stuff was, along with my school records.

THE DECISION

After this incident I thought long and hard about my options, which were few. My plan was simple: I would join the Army and stay in for thirty years or die trying. The only downside I could see was not making it to the thirty-year point.

I told my mother about my new plan, thinking she would be happy, but it was one of those awkward moments. She asked a number of questions about how I'd made this decision. As I told her she kept asking, "Did you think about this or that?'' Her main points were that I did not respect authority, did not play well with others, had no control over my temper, etc., etc.

I explained that I *wanted* to be an infantry paratrooper in the U.S. Army which means that someone would be teaching me something I wanted to learn. I told her they had a delayed entry program so I could join at sixteen. I thought the sooner I got in, the better. She started to cry and said, "You have always chosen the hardest path. On your sixteenth birthday, I will sign the papers."

On my sixteenth birthday I went down to the recruiting office and told them I wanted to become an infantry paratrooper. The sergeant asked my age and I lied and told him I was seventeen. He told me that if I could wait about an hour, I could take the test today. There were maybe five other guys taking it, and they all looked a lot older than me.

The sergeant told me that if I passed, they'd mail me a time and date to take the physical. He gave me an enlistment contract which explained the agreement between the Army representative and me.

The contract specified that I would be trained as an infantryman and go to jump school, with a provision that I meet the prerequisite standards. Failure on my part would not release me from my three-year commitment, but then my role would be determined by the needs of the Army. He told me, "This is an important document." Then he said, "You should keep a copy of this with you, in case a problem comes up." I folded it and put it in my wallet.

I do not think you could study for the tests they gave and I am not sure how they work. But they look for personality traits, problem-solving skills and other things. The test asked things like, "Would you rather be a violinist, a school teacher or a lumberjack?" I was honestly worried about what the questions were asking. It was about two or three hours, as I recall. When I walked out, I was pretty sure I had failed every test, but I took the physical a few weeks later. Then there was nothing for three weeks.

I was sure my plan was in the wind when one evening an Army sergeant first class (SFC) in dress greens knocked on the door asking if he could talk with my parents and me. I thought I must have broken some law.

He was very nice and explained that they regularly delay entry but that I must have made a mistake on my birth date. I would have to wait until next March to enlist.

I thought I was in trouble with my mother when she said, "I'm sorry, he's just in a hurry to jump out of planes and get himself killed." The sergeant was very nice and acted like it happened all the time. I remember thinking that military guys are just like they are in the movies. I liked that they were nice to people when they could be.

He went on to explain that I had passed the exams and the physical. I was cleared for enlistment as soon as I turned seventeen.

MELODYE

I was about sixteen when I met my first wife, Melodye. She started at the nursing home, working as a nurse's aide, about a year after I did. Melodye's mother had gotten her the job. I did not know her age, but I knew she was older than me. We started off just saying good morning. That was about it for a month or so.

Then she and two other women moved into the apartment a few doors down from my mother and me and I started bumping into them on a more regular basis. I do not remember the other girl's names, but they were all about the same age, just out of high school.

When the girls were unloading a trailer of furniture, I asked if I could help. They said yes, thank you. I helped them unload, which took a while. They had a lot of furniture, large chairs, beds, lamps and a lot of boxes. After the last box, one of them offered me a Coke. I sat and drank it as they unpacked and moved their stuff around. I said thanks and left.

Melodye seemed a little shy and I was not really shy, but indifferent, about females. I was poor and a little standoffish. She became a little more friendly as time went by, asking me to come in and have a soda now and then. I, on the other hand, never asked her to come into my mother's apartment.

I would sit in the girls' living room and mostly listen to their stories. They had all recently graduated from high school. The three of them were sharing the two-bedroom apartment. I really do not remember their names. The tallest one had short black hair, the other one was shorter with reddish brown hair. Melodye was always nice and polite. The other two seemed to be more like my sisters, always trying to get a rise outta me.

I was used to ignoring this stuff and it seemed to bother Melodye more than it did me. They would say things like, "Where is your girlfriend? Do you like girls? Etc." This kinda boiled over one afternoon when Melodye and I were sitting in the living room. As usual, I was mostly listening when the tall one sat in my lap and started mussing my hair. It really was no big thing. But Melodye got angry with her and as they discussed it, I left.

I really do not remember how long it was before she asked me to come in and have a coffee. I declined, saying I didn't think her roommates liked me hanging around and I was way too young and poor to be a boyfriend. I think this hurt her feelings. We went back to saying good morning and that was about it for a while.

I thought she was very attractive, but I was in no position to do anything about it. One day the tall one asked if I would help her move some furniture. I said, "Sure, when?" "Now," she said. She wanted to move the two twin beds about 90 degrees. It took ten minutes. When I started to leave, she put her arms around my neck and kissed me. I didn't kiss back. I just said, "I gotta go and thanks for the kiss."

One night after that, I was going into my mother's apartment when Melodye was coming home and asked me to join them for some chicken. I was sitting in the living room with Melodye and the tall one came out with nothing on but panties. Melodye was pissed. She asked her roommate what she was doing and the tall one said, "It's okay, he's gay." They both looked at me. I was smiling.

I said, "I'm not gay, I'm poor." She got red all over and went into her room. Melodye looked at me and I said, in my own defense, "She does have impressive breasts." Then I got up and left. No chicken for me, I thought.

There was a serious change after that. The tall one started smiling whenever she saw me. I got the idea she was flirting. I think Melodye got the same impression because she started coming over to my mom's apartment when I was alone and my mother was at work. We made out but that was as far as it went.

One day Melodye came over and told me she and the short one were going to St Louis with her boyfriend and that he would get us all jobs at his uncle's restaurant. She invited me to join them.

I had just turned seventeen about two days prior and was all set to join the Army, but St. Louis sounded like a good idea. It wasn't.

St. Louis

After we got to St. Louis. The boyfriend took off. The three of us were stuck. Getting a job was impossible.

I thought about the master sergeant who'd come to visit me and my mother. He'd told us I couldn't join the Army until I turned seventeen, but my paperwork was on file and if I decided to go, I was already approved.

After a couple of days in St. Louis, I told Melodye I was going to enlist. I went to the recruiter and gave him my information and told him to contact the Denver office for my paperwork. He said it would take a couple of hours and if the paperwork was in order, I could leave that evening. It was and I did.

I gave Melodye what money I had so she could get a bus ticket back to Denver.

I got a bus ride to Ft. Leonard Wood in Missouri and started my career in the Army.

After jump school I went back to Denver while I was on leave and Melodye and I got together. I went back to Ft. Bragg, and then got deployed to the Dominican Republic with 7th SFG.[26] I told them to hold my mail. Everyone said we would be gone for a couple of weeks at the most.[27]

When I got back, Melodye had written to me to tell me she was pregnant and had given the baby up for adoption, which my mother and sisters weren't pleased about. I married her about seven months later when I was on leave after I re-enlisted to go to Vietnam.

I re-enlisted to go to the 173rd Airborne Brigade in Vietnam, got a thirty-day leave and went to Fort Dodge, Iowa, to convince her to marry me and meet her father. She said yes and we were married in Denver I think in late '67. Then I was off to sunny Viet Nam. I was eighteen and she was nineteen or twenty.

When I got back, we traveled Europe in a Lotus Elan, a Porsche and a Jaguar. I took her to Paris and we toured Germany, Switzerland, the southern wine district of France, England, Ireland and Scotland. I put her through college and most of grad school. She had two miscarriages but we never had any other children.

We'd been married about seventeen years when she told me she wanted a divorce. I went to the senior warrant officer course at Ft. Rucker for six months and she fell in love with her boss. We had not been getting along, partially because of my job, and partly because she had significant issues with men in general. It seemed like going for her master's degree in psychology reawakened a lot of her issues. We were divorced in about six weeks. It was not a contentious divorce.

---

[26] Special Forces Group

[27] He was gone for eleven months.

Part II: My Military Career

This cannot be all inclusive. I've included what I think were the significant events.

BASIC

You already know that I joined the service from St. Louis, Missouri, on April 17, 1965.[28]

Out of money and ideas, I simply walked into a recruiting office and explained that I was already cleared for joining the Army. I was sworn in and on a bus to Ft. Leonard Wood, Missouri, that evening.

There were about twelve guys on the bus and when we got to Ft. Leonard Wood around nine or ten that night, there were more buses. In all I would guess there were maybe fifty to sixty recruits. Everyone had a suitcase except for me. I had a jacket and sixty-eight cents in my pocket.

We were led into a large wooden building and into a room that was very brightly lit. School desks were arranged in neat rows, all perfectly aligned. I would guess there was seating for maybe eighty to a hundred people.

One of the soldiers told us to take our suitcases to the back of the room and then be seated. As I had no suitcase, I sat in the first front row seat. As the others came to sit down, they spread out. A few sat in the back row with empty seats between them. When I looked at how we were spread out, I thought, "I hope we don't have to pass stuff out."

We sat there for about twenty minutes in absolute silence. The soldiers who had brought us up the stairs were standing by the door. Suddenly the door flung open and a very tall, very black and very muscular sergeant first class walked up to the platform at the front of the room.

He seemed to look directly into each of our eyes. Then, in a voice that could have been heard clearly in St. Louis, he said, "Do any of you fucking recruits know what the fuck is in store for you?" Quiet. "Do you?" Silence.

He laughed a loud, deep laugh. Then he said, "Your lives are about to change."

He asked if there was anyone in the room who felt they could whip his ass. Again, no response. "How about you boys in the back, any takers?" No response.

He then said, in a normal conversational tone, "Get up and line up against the two side walls." Of course, the guys in the back did not hear him and stayed in their seats, or stood up but did not know what to do.

---

28 My father was seventeen years and twenty-two days old.

In his command voice he said, "I am the BBTMFIC. That means, the Big, Black, Tall, Motherfucker In Charge. So get your puny little asses against the fucking wall or come up here and kick my ass. NOW."

The result was very impressive.

"There are many ways to do things," he said. "The civilian, half-ass, when-I-have-time-to-do-it way. And then there is the right-now, no-hesitation, get-it-fucking-done-and-done-right Army way. And you are going to learn, and I mean right now, the Army way. So, when I say go, you will file in an orderly manner and fill the seats from the inside toward the wall. Every fucking seat. The desks will not move from their current position, you will not say one fucking word. You will fold your hands on top of the desk. You will sit straight up like fucking men. Are there any questions?"

There were none. Once we were re-seated, it was like a Jekyll and Hyde thing. He started speaking in a normal voice and explained the sequence of events over the next few days.

I was impressed with the way he took command and had us all following his lead in five or six minutes. He introduced a personnel sergeant who gave us the instructions on filling out each form. We spent two hours or so doing that.

Then we were all issued toiletries, sheets, two blankets, one pillow with a case, and escorted to a barracks. We were told we would be there for the next three or four days. There was little talk and we were all soon fast asleep.

My first morning a sergeant walked in the barracks and told us we had a very busy day and to get dressed and to fall out. He then explained that that means to get yourselves outside.

It was a very busy day. We were given forty-five minutes to go through the breakfast line, eat, drop our trays off and fall back outside, where he would be waiting. The first stop was the barber shop, where we got buzz cuts. Half of the hair went down your collar. It was not pretty, but it was efficient.

Then it was time to draw all of our uniforms. We lined up and went through in a single file. We put our feet on the footprints painted on the floor and stood straight. They measured us, asked our shoe size, and told us to follow the yellow line. The first thing I was given was a duffel bag and was told to put everything I was about to be issued into it—except the boots, shoes and bus driver hat.

The efficiency and speed with which we went through that line was absolutely astounding to me.

I do not know if anyone has ever filmed this, but if so, I would love to see it. Not the crap you see in movies where guys are wise-cracking, because that was not the way it was. I mean, we went through that issue at almost jogging speed. We signed for whatever they gave us. Then went back to where we were staying with two pairs of boots around our necks, low quarters (shoes) in one hand, the duffel bag over our shoulders and a bus driver hat on our heads. We did not know how to march yet, so we just walked in a column of fours.

We went to our beds and the sergeant said, "Gentlemen, listen up. You have just signed paperwork that says you received (some dollar amount) of clothing. This is the one time in your military career that you will have the opportunity to check after you have left the issue point to make sure you have exactly what you are supposed to have. This will never happen again in your career."

He went on, "I have had photos of all the items you were issued put up on the walls with the name that is on the paperwork receipt you were given. Prior to 06:30 tomorrow take the time to check each item against your equipment and the list. If you are missing anything, tell me tomorrow. After that if you are missing anything, it will be deducted from your pay. Are there any questions?" There were none.

I think it was about then I realized these Army guys did not play. I told myself to listen closely and if I had any questions to, by God, ask them. That self-given advice has saved my ass, and the asses of others, on many occasions.

If you ever think, "They will probably cover this or that later," don't. If you have a question or are not absolutely clear on anything—ask. And if the answer is not clear, and I mean crystal clear, repeat the question until one of two things happen. One, you're clear on what you are being asked to do. Or two, you are scrubbed from the task at hand.

The next few days there was more testing. I remember thinking, does this testing shit ever stop? Short answer: not in the Army. I am not sure if it was two or three full days of testing, physical exams, written tests on topics undisclosed, personality traits, IQ testing… you name the test, we took it, or one like it.

I think we had been in processing for about four days when a specialist came and said, "If I call your name, please fall outside and wait for me." He called about five or six names, including mine. When he came outside, he told us to follow him.

He took us to an old wooden building with a long hall and seating on both sides and told us to take a seat. Then he said, "You will be called in one at a time. When your name is called, walk to that door at the end of the hall, knock, and wait to be told to enter. Are there any questions?" There were no questions.

The others looked worried and I started to feel uneasy. It took maybe twenty minutes before a private first class (PFC) came out from a side door, called a name and pointed to the door at the end of the hall. The guy walked to the door, knocked, and was told to enter. He never came out. Twenty or thirty minutes later the PFC called another name, pointed at the door and went back through the door he had come from. I still wanted to know where the first guy went.

They had been calling us alphabetically for everything and we had not been called or sent anywhere alone since we arrived. Now we were being called seemingly at random, one by one going into a room and not coming out the door we'd entered.

There were about nine or ten of us sitting in that hall. I admit that when I was the only one left waiting, I did not feel very comfortable. Finally, my name was called. I knocked at the door. A voice said, "Enter." I walked in and remained standing. An officer sitting behind the desk looked up and said, "Take a seat." I sat in the only chair available. He was reading from some folder for a few minutes, then he looked at me and smiled. "What do you think of the Army so far?" I answered, "It seems to be far better organized than the places I've worked or any of the schools I went to." He laughed. I did not. I still felt like a hammer was hanging over my head.

I said, "Sir, may I ask a question?" He said I could. "What is this about, why am I and the others here?" "You are here to be offered a rare opportunity. The others were here because they failed one or more of the tests or they falsified information on their applications. They were given the opportunity to retract their enlistment or be court-martialed for their false statements and sent to prison."

"Sir, may I ask what the opportunity is?"

He gave me a friendly smile. "All the tests you've taken give the Army a lot of information. In your case it tells us you have above-average intelligence and excellent hand-eye coordination. You are an independent thinker, and you are not ashamed of your mistakes. You learn from them."

He offered me, at the Department of the Army's request, an opportunity to go to OCS, Officer Candidate School, then to a university, or to flight school with no university. I asked more questions and he answered those he could.

I sat there thinking for some time. Then I asked, "Do I have any say in this?" He told me it would be my choice whether to accept, or to just do the three years I had signed up for.

I explained that I was only seventeen and had a lot to learn about the Army, and myself, before I would be comfortable giving orders to others. I asked, "Can I still join if I just want to be a paratrooper?"

At that moment I wanted more than anything, to become an infantry paratrooper. And that was my focus. I knew I was in for a lot of shit and told myself to keep my mouth shut, my eyes and ears open, and my focus on the goal. I think if my mother had been there, she would have told me the same thing.

He smiled again. "This is not a one-time offer. This information is a permanent part of your record, so if you do apply at some future date, you will be accepted."

Then he said, "Young man, you are wise beyond your years, and I would like to shake your hand and wish you good luck in your Army career."

I asked one more question as we shook hands. "Will my commanders have to know about these tests?" He said they could ask for my records and were entitled to them on request, but he didn't think any commander had the time to read the results of all their soldiers' tests. I left the room the same way I came in. When I opened the door, the sergeant was waiting to lead me back to my barracks.

The barracks was empty. All the beds were clean, save mine. I had the place to myself. I went to dinner and came back to find the place still empty, save for me. I didn't know any of the guys and hadn't introduced myself to anyone, but I still felt like I had been abandoned.

The next morning, I was awakened by our guide and told that I would be picked up in a couple of hours to start basic training so I should get breakfast. After I ate, I came back to sit on my bunk and wonder if I had made a very big mistake.

When the guide came back, he asked if I wanted him to show me how to pack all my gear in the duffle bag. He showed me how to fold and where to place my stuff. He also explained how to unpack it and put it all in my locker when I got to my barracks for basic. He was very thorough. I practiced folding and rolling my stuff quite a few times.

Before lunch a sergeant came in and asked if I was Bailey, Joseph. I said I was, and he said, "Grab your kit. I'll take you to your company." I was a little overwhelmed when we pulled up in front of what looked to be a brand-new, four-story brick building. There were four of them in a row and four more across the street. They were huge, each one a city block-long. The sergeant pointed and said, "That's your door. Good luck."

I stood there wondering where everybody was. The streets were empty, no one was visible in any direction. Inside the place was spotless. There were a couple of what looked like offices and a room full of Coke and candy machines, with maybe four pool tables. (Which none of the trainees were allowed to use, ever.) It all looked brand new.

I laid my duffle down and just stood there. It may have been ten or fifteen minutes before someone came down the hall and asked who I was and what I was there for. I gave him my name, told him I had just finished my in processing, and was supposed to start basic training. He told me I was a bit early and that he would tell the first sergeant I was here.

Twenty minutes later a drill instructor took me to the fourth floor and told me to pick a room and settle in. "Staff Sergeant Parker will be here on Thursday. Until then you're on your own. The CQ will wake you at zero five thirty tomorrow. I will have someone show you where the mess is." Then he was gone.

What was a CQ, what was Zero Five Thirty and what kind of mess was I supposed to clean up tomorrow?

I picked a bed and made it. I opened the clothes closet and there was a chart of how everything was supposed to be put in it. I gave myself a tour of the building. All the rooms were the same, four beds with a closet and two drawers. The toilet and shower areas were huge. I do not remember how many shower stalls, toilet stalls and urinals there were, but I do remember wondering about how many guys they were expecting.

As promised, the CQ (Charge of Quarters) showed me where the mess hall (dining facility) was. Which was directly behind the barracks, also absolutely huge and three-quarters full of new recruits. I grabbed a tray and walked through the line. Then I stood there for a moment. Everyone was eating as though they were starving. I walked to an empty table.

When a table of guys went out, they marched. I thought they looked pretty good.

When I got back to the barracks, Staff Sergeant (SSG) Parker was standing on the entry steps. He said, "You Bailey?" I told him I was. "Can you drive?" "I can, but—," "No buts, can you, or can't you drive?" "Yes sir, I can," I answered.

He looked at me and said, slowly and very sarcastically, "I am a staff sergeant, I work for a living. You only call officers sir. Now get in the car." We drove somewhere. When he parked, I followed him into an old wooden building and up to a desk where another sergeant was seated. They were obviously friends.

The gist was that I was about to take an exam to be a licensed driver. When he turned toward me, I said, "Sergeant, I am only seventeen." He said, "A soldier's age is not important. Go with him, I will pick you up in about two hours."

The other sergeant gave me an eye exam then a sign recognition test. Then we went out to a huge Army truck. I said I had never driven anything like that before. He—honest to God—slapped me on the back and said, "Driving is driving. You'll get used to it."

I got in. I looked at all the switches. I was definitely out of my comfort zone. He talked me through the starting procedures and said, "OK, let's go." I put the truck in gear, gave it some gas and we were off. He had me drive to what I can only describe as an obstacle course. It was one kind of obstacle after another. The instructions were to stay on the course to the end. He was making marks on a clipboard as I drove.

When we got to the end there was a very large painted sign that read *The End.* Then he said, "Back to the motor pool." I waited for more instructions but none were forthcoming. He looked over at me and asked, "What are you waiting for?" I said, "I have no idea what or where the motor pool is." "How long have you been here?" he asked. "I got here yesterday." "Oh!" he said, and directed me back to where we started.

When we went inside, he told me to have a seat. I sat in the hallway for a few minutes until he came out and said, "Here it is, your license. SSG Parker will pick you up in a bit." I was rather pleased with myself for getting my driver's license a year prior to my contemporaries.

SSG Parker drove up and I got in the car. We did not speak. He drove me to our unit motor pool and introduced me to two noncommissioned officers (NCOs) and told them I would be their platoon driver. They said, "Welcome to the Army, Private Bailey." I said thank you. They made a copy of my license and we drove back to the barracks. As we drove, SSG Parker said to pay attention how to get back. I did.

He dropped me off in front of the barracks and told me to walk back to the motor pool, then drove off. When I got there, he was there with two sergeants. "I told you he was sharp," he said to them. "Let's go back to the company."

New trainees started coming in but none on the fourth floor where I was. Then a bunch of buses pulled up out front and within a couple of hours we had a full house. I was sitting on my bunk when three guys walked in who all looked like they were in their mid-twenties. Walter Verstrate was from a small town in Oregon and his family had a funeral home business. Arthur McArthur was a semi-pro boxer from Chicago. The last one seldom talked and kept to himself. I don't remember his name. I introduced myself and offered my assistance getting their lockers set up.

The next morning was the start of a new life for all of us.

We were an unusual mix of young men from all strata of wealth, education, and ancestry. We descended from immigrants and natives from every race on earth, a mix of volunteers and those who had been drafted. There was no segregation in the Army. Nobody gave a shit who your parents were. We only cared about each other's ability to be part of the team.

The first morning we fell out for formation, SSG Parker inspected each of us, making comments about how we were to have our uniforms.

When he came to me, he asked, "Why didn't you shave today?" I explained that I was only seventeen and wasn't shaving yet. He explained that all soldiers shave every morning. I told him I didn't have a razor. He reached into his pocket and handed me one. He then explained how to use it and had me demonstrate for the entire platoon. I learned two things that morning: that rules of conduct and uniform appearance are not suggestions, and that shaving without the benefit of water, shaving cream, or a sharp razor blade is not something I wanted to ever repeat.

Looking back, it still amazes me how they took us from our civilian mindset to speaking, thinking and living together as a team. No, it was more than that: we became a family unit. We watched out for each other and helped one another meet whatever challenge they presented us with.

As one of the platoon's drivers, I got up early to get a truck whenever we were going to a range that would take too long to march to. I would park it out front and go to breakfast. On one particular morning, when I walked into the office of the motor pool, two NCOs were listening to the radio. They both held up their hands to indicate I should be quiet. So I listened to a newscaster reporting that we (the U.S. government) were sending the 173rd Airborne Brigade to Vietnam. The 173rd Airborne Brigade was trained, equipped and focused for one purpose: to close with, kill, and or destroy the opposition's ability to resist.

This was a new step. Regular Army ground combat forces were going to Vietnam.[29]

The officers' response to each other was kind of alarming. They said, "We need to decide to retire or to get in shape. We are going to war." Then they looked at me and said, "Private, your timing may be off a little."

I never mentioned any of what I heard to anyone. I figured it was the NCOs' job to tell us. They never said a word. I do not remember anyone saying anything about it. It just kinda happened. I think people outside the military were sorry about the monks setting themselves ablaze, but I do not think anyone outside of Washington and the military had a clue as to how deeply committed we were or how long we had already been having our soldiers killed there.

I was totally ignorant, as I was with most of what was going on in the world. One of my many character flaws is if I am not in some way responsible for it (i.e., my family, friends, those I have responsibility for, like men under my command) I don't care about it. I know that makes me a bad person to many people. But if I am powerless to change, stop, avenge or correct a problem, I have never understood why I should concern myself with it. I mean people getting all depressed about the weather, or a forest fire on the other coast, it's never made sense to me. I really don't give a shit about it. I am just not wired that way.

It was in those last few weeks of basic training that I wrestled with the question of whether I had the nerve to put myself in harm's way. I decided that I did.

---

[29]This was 1965, the year the US went from less than 25,000 troops to over 150,000 troops in country.

Jump School

About a month in, SSG Parker called out three names from the entire company and told us to fall out and report to a sergeant at the rear of our formation for Jump School. It was Walter, Arthur and me. The sergeant told us to get on a waiting bus. We arrived at a physical training (PT) field. There were about nine buses with around sixty or so soldiers, all still in basic training.

Waiting for us were five or six NCOs dressed in fatigues, trousers, and white T-shirts with their rank and names printed on them. There were also NCOs wearing a different kind of uniform and some green berets. The guys wearing green berets gave us about twenty minutes of what they did and said, if we wanted to be one of the Army's best, we should ask to join them after jump school.

They called our names and put us in four groups. Then we went to different areas. They explained, "All of you have one thing in common: you have requested to become paratroopers. We are all trainers at Ft. Benning, Georgia. We are here to evaluate your physical condition."

We were to be given four physical tests and failure on any one of them would result in denial for jump school training.

I was as worried as everyone else was. We rotated from one test area to the next as a group and were tested in small groups of five or six at a time. They called our names and we repeated our RA, US or NG Number. (RA = Regular Army, US = Drafted for service, and NG = National Guard or Reservist.)

The physical tests were push-ups, sit-ups, and squat jumps, followed immediately by a thirty- or forty-meter sprint. This was, in my opinion, designed to see if you really wanted to be a paratrooper, because it was very painful. After doing the squat-jumps, your legs feel like they are on fire and making them move as fast as you can is like pouring gas on your burning muscles.

A lot of the guys did not finish the run. When they called the names of who made it, some of the guys who didn't, started to cry.

After the test we were dropped off at our barracks. When we got in everyone was still out at the PT field, but almost as soon as we arrived SSG Parker walked in and asked if anyone from our platoon failed the test. Arthur called us to attention and said, loud and clear. "No, Sergeant, not one soldier from our platoon failed the test."

He almost, but not quite, smiled and said, "Take a break." We were the only ones from the company who'd asked for jump school and he was the only drill staff sergeant who wore jump wings. We were all airborne volunteers.

Arthur, Walter and I planned to get airborne tattoos on our first leave at Ft. Leonard Wood, but when the time came, I was not drunk enough—but Arthur and Walter were.

Following orders, I went home on leave and returned to Ft. Leonard Wood. I had requested, and had been assured, I would go to infantry training and jump school. So I assumed that was what would happen. Never assume.

When we fell out for formation the first morning, we were welcomed to the clerk typist school. The others seemed to be aware of this. I was not. This was not my deal with the recruiter. I raised my hand and the sergeant asked, "What is it?" I said, "I am not going to be a fucking clerk, sir." The moment I said it, I knew it was a major fuck-up. But for Christ's sake, what would a paratrooper do with a typewriter in combat?

The sergeant told me to fall out to the rear of the formation and wait. I did and remained at parade rest for maybe twenty minutes while everyone else marched off. A master sergeant walked in front of me and asked if I thought for one f-ing minute that I had any f-ing choice in my MOS.[30]

Somewhat apprehensive but undeterred, I answered, "My enlistment contract says I can go to infantry and jump school if I keep up my end of the contract, which I have. If the Army is unable or unwilling to uphold their end of the bargain, then I respectfully ask to be released from service."

The sergeant took me to the company commander where I explained that I would rather transfer to the Marines than be a clerk. He took me back to see the first sergeant who, after some threats, took me in to see the section commander.

The commander explained that the Army needed my skills as a clerk. I explained that I had quit school in the eighth grade, had no idea how to operate a typewriter, compose a sentence, or spell worth a shit. He asked the first sergeant to get my records. I reached in my pocket and handed him the copy of my enlistment agreement which, for reasons I cannot explain, I had kept in my wallet. He read it, got up and said he would be back shortly.

When he returned, he was friendly. "It seems we have made a mistake on your advanced training. You have already been scheduled for jump school in four weeks. We are unable to get you through your

---

[30] Military Occupational Specialty

infantry MOS (Military Occupational Skill) in time. Which could delay your parachute training for up to eight months. We can put you in a wheeled driver training program, which will end in time to make paratrooper training. You will not go to clerk typist school. Will that work for you?"

My head was spinning. I had always thought every soldier went to some kind of combat course like demo, artillery, tankers, etc., and if they were hurt, or if the Army needed a clerk, supply, intel, etc., then one would be trained from the unit. I had no idea a soldier could work as a typist, driver, or supply guy for thirty years.

As I was processing this, the captain asked, "Can we work this out so you can go to your parachute training?" I was absolutely dumbfounded. I'd been angry when I ended up in clerk typist school and had kinda lost my mind for a while, but here was this captain sucking up to me.

I asked, "If I go to this driver training will I be a driver for my Army career?"

"Not necessarily."

"Then I'll go, if you get me a slot after I complete this course and jump school to go to special forces." I did not really know what special forces were, but had heard they were airborne elite forces. He looked at me and said, "I will do my best."

Then he handed the first sergeant my new orders.

Driver training was not difficult. I learned to drive everything from a Jeep to a five-ton truck.

Ft. Benning, Georgia

I went to jump school at Ft. Benning, Georgia. To my surprise and delight, Arthur, Walter and I were all in the same class.

We arrived in June or July; I remember it was hot. Ft. Benning has more schools than just paratrooper training, but jump school was the last stop on the bus. The harassment began upon arrival. It was dark and the cadre were, as predicted, hard asses. We got our bags and they double-timed us with our duffels on our shoulders up and down a hill and eventually to our home for the next three weeks. We finally got to our barracks about two hours later, exhausted.

We were all settling in when four NCOs dressed like the drill instructors who'd met our bus came into our barracks. The one wearing master sergeant stripes explained there had been a lot of thefts and that they were there to collect what we did not want to get stolen. He asked everyone to drop their valuables into—I'm not exaggerating—the pillowcases his assistants were bringing around. Alarms went off in my head. This was bullshit.

I was the only slick sleeve in the group, meaning I had never been to combat. I was a private E-2. We had master sergeants, a bunch of buck sergeants, and the rest were specialist 4s. I thought these ranks demonstrated some amount of smarts, but I was wrong. As God is my witness, these full grown, experienced combat soldiers took off their watches, rings, gold and silver necklaces and put them into a pillowcase. The one wearing the master sergeant rank assured them that they would get everything back at morning formation. When they got to me, I said, "I have nothing of value." They simply kept going down the line.

I figured maybe these guys weren't thinking straight. We were all tired and about to try and become paratroopers, which can be very intimidating.

The next morning, I waited for someone to ask for their stuff. It took maybe ten minutes. At first the sergeant thought we were kidding, but when the first sergeant asked for a show of hands of how many had given their valuables to the men posing as cadre, all but seven of us raised their hands. Everyone who raised their hand was asked to fall out.

I knew then that the insignia of rank on a soldier's arm, or the insignia of rank on an officer's collar, did not give any indication as to their intelligence, only their pay grade. This has been the source of many disputes.

The rest of us were told to fall in with another group and we marched off to what was called the PT Pit. It was a square, fifty yards by fifty yards, maybe two feet tall and filled with sand. We were then led in physical drills for what seemed hours.

The assistant drill instructors were walking around the pit correcting our form when one of them shouted, "I got one." Then another shouted, "I have one." This was followed by the command to recover, which normally meant we should stand up and wait for the next exercise to be demonstrated.

But this was not that. The drill instructor told his assistants to please bring the guilty parties forward, so we could all get a look at them. I thought, "Wow, they caught the thieves already." But it was not that. They had found about seven or eight guys with airborne tattoos, including Arthur and Walter, but none of us had our wings yet.

It would be pretty embarrassing to have an airborne tattoo without graduating. But the cadre convinced two of them to resign right there. I made a mental note to never say the words, "I quit."

After the morning drill we were marched to the training field and told to fall out and form a circle around the instructor. I took this opportunity to go find Arthur and Walter. They grinned when they saw me and said, "You were right about getting the tattoos."

They were in different barracks but we did most of our training as a trio when we could. The morning run was brutal. We had our calisthenics after breakfast and then our run. I have no idea about the distance, it seemed to depend on the instructor's mood.

We would start at a double time, and around the three-quarter point we would start running full tilt. Two things would happen: first, our formation went to shit. Second, the fastest runners would move to the lead and the main group would be four or five times as long, from the front runner to the last. They harassed the slow guys, and sometimes failed them. We had at least one guy every day that just said, "Fuck it, I quit." I think that was the idea all along.

The morning run during tower week was my weak spot. On the first run in the third week, I was one of the slow guys and fell back about thirty or forty yards all by myself. There were maybe three guys behind me, but I never looked back.

One of the drill instructors fell out of formation to wait for me. Sure enough, he started his spiel, "Kid, you don't need this. Nobody is gonna care if you just quit. Give it up, you'll feel better." I was trying to close the distance with the group and was now about fifty to sixty yards behind. Then they all suddenly stopped; they had completed the run.

I picked up my cadence and was sucking air through every orifice I could. Then here came Arthur and Walter, running toward me and screaming, "You can do it, come on." They were next to me, running with me, and saying silly shit like, "You can't help having short legs." The DI told them to get back with their group. I told them, "I'll make it, go ahead." They said, "Shit, we ain't running any faster, we're tired." I was the last one to complete the run, but I completed it. The ones behind me were never seen again.

For the rest of the week, Walter and Arthur ran with me. I felt like I was holding them back and said so. They told me they were running with me because they wanted to. I think those were the first (and last) guys I ever thought of as being the kind of man, and soldier, I hoped to be. They were like big brothers to me.

I was the youngest guy there. I looked like a kid playing soldier.

I got faster and by the weekend we all stayed with the main group. At the end of every day's run, we would line up and walk through the showers and be given two salt tablets to swallow. I could not see myself, but all of us looked like we had been dunked in boiling water, our faces and arms were bright pink. We were allowed maybe twenty-five minutes to catch our breath. Then back for formation and parachute training.

They had a track that circled the training area. Inside this was the Parachute Landing Falls, or PLF, towers. Next to the towers was the suspended agony shed. They were covered so, rain or shine, we could practice our parachute landing falls. This was everyone's least favorite exercise. They had us buckle into a parachute harness and then hoisted you just off the ground. We called the harness the ball buster for obvious reasons. We practiced checking our chutes, making sure they were opened and properly deployed. We practiced bicycling, a maneuver to untwist the static lines so you could steer the parachute, in case of a Mae West (which is when one or more of the risers cut across the chute and it resembles a bra) or streamer (a parachute that has wrapped around itself and cannot open). Then we would practice the landing position.

At first, we jumped from three-foot platforms, then from six- or seven-foot platforms. It was practice until tower week.

Tower Weeks

There were two tower weeks. First was the thirty-four-foot tower, which is designed to test your fear of heights. Thirty-four feet is the height at which you can visually recognize how far above ground you actually are. They put you in the harness before you climb the stairs. Then they give you a helmet with a number painted on it for a cadre on the ground to watch to see if you shut your eyes and if you remember to check your chute. You are attached to a cable and a large bunker with a stop cable on it, but guys have to grab you to keep you from sliding back. It's easy to get stuck hanging. If you did not do the jump correctly, you were told your mistakes and did it again until you got it right, were dropped from the course, or quit. I was kinda shocked when I passed on the first go.

The second tower week is the one prior to our first actual jump. The towers are two hundred feet high and the parachutes are attached to a rig that opens them prior to being raised. One by one, we were strapped in and then slowly raised. Just before dropping us, they would ask, "Are you ready?" This was another opportunity to quit. It was a little intimidating, hanging there two hundred feet in the air. It was the scariest part of ground school and it is where most guys quit.

When we arrived at the bottom of the tower, it certainly seemed much higher than it had from afar. We were all staring up at it and suddenly an instructor we had not seen before told us through a bullhorn to sit down and listen up. We all sat, and he proceeded to tell us what we had been doing and what we were about to do. Then he said, "Listen carefully. If you have ever found yourself looking down from a high place and had a feeling of weakness in your knees or legs, felt slightly nauseous or very uncomfortable, raise your hand and keep it up." Three or four guys raised their hands. I wondered why in the world would someone with a fear of heights volunteer for jump school, and I still have no clue.

There were three or four catches that would be lowered from the top and a soldier would stand ready to hook the chute to the system. The chute is completely deployed inside the hoop then they are raised about twenty feet in the air. You get under it and strap yourself into the harness. They tell you not to steer toward the tower. Duh. Once you're strapped in, the instructor checks that you're hooked up. Then the lead instructor asks if you are ready. If you do not immediately answer in the affirmative, you are taken out and go to the back of the line. The controller raises you to the top. There is a short pause. Then you feel a jerk and hear a loud click. There is only one way down from that point. When the instructor asks if you're ready, you must sound off loud and clear, "Airborne." You are graded on doing what you have been trained to do. Check your canopy, control your direction of descent, execute a good parachute landing fall. You only get two chances. If you do not complete the sequence correctly, you fail the course.

We were all a little worried about the tower. A few guys lost it and screamed on the way up or down. The ground below was plowed and soft. When I was dropped, it was really kinda neat floating down.

The ride up was kinda like an amusement park ride. The sudden stop was a little exciting. Then the click was very loud. This indicated all safeties were disconnected, so that when the button was pushed you would be in free fall. The ride was short and sweet. After I stood up, the instructor gave me the thumbs-up. I only had to do it once. I was going to Jump week.

BOXING MATCH

Saturday we were awakened in a nicer manner than the normal banging on trash cans. It was actually a little scary to wonder why they'd suddenly stopped being such hard asses. "How many guys died last week?" I asked, and everyone froze and looked terrified. Then one of them laughed and we all did. But we still wondered.

After breakfast, most of the guys wrote letters or read magazines. I took a nap for an hour or so. It was awkward not to be doing something.

We spent the Saturday before the real jump at an all-comers outdoor boxing match. I think there were six or eight matches scheduled. Walter told me Arthur had signed up to fight.

Walter said he felt sorry for the other guy, whoever he was. Arthur's match would be the last one, so he came and sat down to watch for a while.

It was definitely entertaining, but it wasn't like television boxing. These guys were not very good, but they fought hard. When it was time for Arthur to leave, I said to Walter, "I bet he wins." And some guys behind us said they would take that bet. I had fifteen dollars and bet all of it. Walter went nuts and bet a lot more. I said, "I hope you have the money, 'cause I bet all I have."

Out came Arthur and his opponent. The guy was quite a bit larger than Arthur and looked like one of those body builders. I looked at Walter and said, "They can't be in the same weight class." "I don't think they have weight classes here," Walter said.

It was the mad bull versus the cool, calm, and collected matador. Arthur's head was in constant motion, as was the rest of him. The only target he seemed to aim for was the guy's face. After the first round, the poor guy looked like he had been hit in the face with a meat tenderizer. I never saw him actually connect any of his punches. About halfway through the three rounds, the cadre stopped the fight. Walter and I were on our feet screaming, then everyone was screaming.

We were paid and we both gave Arthur half of our winnings. He wasn't even sweating. I asked if the guy was above his weight class and he said, "Weight is not a factor unless both opponents know how to box."

Just about everyone went to the beer hall and drank pitchers for a couple of hours. That was the one and only time we were allowed to go to the beer hall and have any kind of recreation. I think it was to give us something else to think about besides next week's jumps.

Sunday was low-key. But Monday was back to normal. Up and at 'em. We fell out and for what I think was the first time all the different chalks fell in and did our run together. When the drill instructor gave the preparatory command of double time, we all responded, "Uh-rah, airborne." And off we went for the first day of the last week in jump school. I know most people would not understand. But for us—and for those who went before and after us—this week would always stay with us.

THE JUMP

Once we'd jumped three times, we were paratroopers. Our final training jumps were from twelve hundred feet. That is a significant difference from thirty-four feet or two hundred fifty feet.

My first jump was the first time I was ever in an airplane. We got to the airfield and into our chutes and on to the waiting C-130 Aircraft. It was exciting. I was thinking, "If I don't fuck up and kill myself, I bet my Mother will be proud to have a paratrooper for a son."

No one talked. It seemed like a library, except for the aircraft engines.

Then they opened the doors. It was extremely loud and made you wonder what in the hell you were about to do.

Above all that noise, we could not hear, but we could see the hand and arm signals. Unfasten seat belts, stand up, hook up, check equipment, sound off with equipment check, etc.

Then finally came the command "stand in the door." The first man in line would literally step into the open door and wait to feel the slap on his shoulder signaling him to jump. Then we all followed as quickly as possible. When we jumped, we tucked our chins, crossed our arms over our reserve chute, counted one, two, three, four—and hoped we felt our chute deploy. Then we looked up to check and confirm it had properly opened. No streamers or Mae Wests. Both had to be addressed immediately for any chance of survival.

The time aloft was not very long, but for me the first jump was like winning gold in the Olympics. The next two jumps were great too. A few guys quit after the first jump, which I have never figured out.

Immediately after our third jump, we got on buses and were taken back to the two-hundred-fifty-foot tower area. We marched in formation, fifty or sixty guys, three ranks deep. Lined up in front of us were the cadre from day one to our last jump.

Centered in front of them was a colonel wearing a fatigue uniform with master jump wings and a CIB[31] with a star above it, indicating he had served in two separate wars as an infantryman. His white hair was cut close. The captain called us to attention then the colonel called the cadre to attention and gave the hand salute. They were saluting us, and we returned it.

The colonel told us we were now part of an historic combat team and we should be extremely proud of our achievement. He said, "This class started with some number of trainees and you few are the only ones left.

---

[31]Combat Infantryman Badge

Turn around and look at your numbers." Then he gave us the starting number and the graduating number. I do not remember them, but I do remember thinking I was glad I was in the graduating group.

Then he called us to attention. He stood in front of the first soldier, pinned his wings on, shook his hand, leaned in and said something no one else could hear. I was the last man in the first rank. When he stood in front of me, he said, "How old are you, soldier?" "Seventeen, sir." "I assume you joined as opposed to being drafted?" "Yes, sir."

Then he said the same thing as he did to everyone else and handed me the rear clips that hold the wings on. He leaned in and said, "Don't twitch or say ouch." As he stepped back, he slapped the wings, pushing the needle-like pins into my chest and said, "These are your blood wings. You are part of the United States Army's elite fighting men. Keep these wings safe and buy another set to wear. They are priceless. And hard earned." Then he shook my hand and moved on.

My chest popped out an inch. I felt I had finally found where I belonged. These were the kind of people I would want to be by my side if and when I ever saw combat.

Special Forces

After the ceremony we marched back to our barracks and packed. Most of us were leaving that afternoon. I said goodbye and good luck to Arthur and Walter.

I was going to Ft. Bragg, North Carolina, assigned to the Special Forces Warfare Center. It is a short bus ride from Ft. Benning to Ft. Bragg.[32] Bragg is home to all the Army's elite combat forces training centers and the 82nd Airborne Division. Just being there was special for me.

I was picked up and taken to the snowbird barracks. (Snowbirds are those who show up without a class starting date. Blackbirds are those who have not successfully completed a training program and are awaiting orders.) There were about fourteen guys in the barracks, ranging in rank from E-2 to specialist E-4. I was the only E-2. There were a few NCOs, but they were in other barracks.

As snowbirds we were sent out every day to some detail or another. I had been there for about a week when I was assigned an office cleaning detail. The sergeant showed me how the buffer worked and off I went to clean the Special Forces headquarters.

---

[32]It's a seven-and-a-half-hour drive.

When we got in, the sergeant told me to not touch anything on the desk. Empty the trash cans, dust the bookcases, and straighten any furniture out of place. I was in my third office when this really old guy came in with a white T-shirt with *Stillwell* stenciled across it. The name rang a bell, but I could not remember where I heard it.

"I am almost done. Do you want me to come back or finish?"

He said to go ahead, then he asked me how it was going.

I said, "So far I'm not impressed with whoever runs the unit."

"Why is that?" he asked.

"I have been here for almost five days and still haven't been assigned."

"What is your name?" I told him.

"Why did you join the Army?"

I told him, "I think I've wanted to be a soldier and defend my country since I was a kid."

"But why?"

"I think it's because of our history, and the movies I've seen. That may sound dumb."

He looked at me and said, "I cannot think of a better reason."

A full colonel came in and said, "Sir, we need to go to a meeting." The man I'd been talking to went to the coat rack and put on his blouse. He was Brigadier General Joseph Stilwell, the son of "Vinegar Joe" Stilwell of WWII fame, and the commander of the special warfare center. I snapped to attention and said, "Sir, I did not know." He cut me off. "It is fine. I am sure you will be starting soon."

I hoped he had not read my name tag.

The Latrine

The next morning, I was told to report to the first sergeant. He said, "You have a special assignment. A sergeant will drop you off and pick you up around 16:00." He drove me to a building and told me to go in and ask for Master Sergeant Bragg.

I went in and there was only one person there, sitting at a desk in a T-shirt, with a large cigar in his mouth. I said, "Excuse me, sir, can you tell me where I can find Master Sergeant Bragg?" He looked at me as though I was from outer space. "Why?" "I was told to report to him, Sir." He got up and said, "Your name Bailey?" "Yes sir." "I am who you are looking for. Come with me."

He led me through a hall and into a latrine. "This is a shit hole, I want you to clean it. Follow me." We continued down the hall through another door that opened to another large room full of shelves and bins.

He introduced me to a specialist 5 and told him to get me whatever I needed. Then he told me to come get him when I was finished.

I asked the specialist if I could have a writing pad and a pen or pencil. He gave them to me and I said I'd need to make a list. He said, rather rudely, "Fine."

Back in the latrine I made my list: bucket, mop, scouring pads, brush, broom straw, dustpan, trash bucket, eight-foot step ladder, four cans scouring powder, rags, steel wool, a can of Brasso, a bottle of window cleaner, and a gallon bottle of bleach.

I gave him my list and asked if he needed any help. He looked at the list and back at me. Then he smiled and said, "Sure, come with me." We got in an Army three-quarter-ton truck and he drove us to pick up all the cleaning stuff, then somewhere else for a bundle of rags and finally to get the stepladder. He helped me unload it.

I started with the ceiling and light fixture which, like everything else, was filthy. Either they wanted me to quit before I started, or they were in severe need of labor. Either way, I was pissed.

I did not have a watch. I felt like if I stopped to eat, I would not want to start again. So I kept cleaning and scrubbing until the ceiling was actually white. The floor was nicely tiled, white and black. And the green pipes for the showers and sinks turned out to be brass. I have no idea how many bags of trash and crap I took out, but it was a lot.

Finally, I was done, or I hoped I was. I took all the extra stuff to the supply specialist and put the leftover cleaning stuff on a shelf in the latrine. Then I went to tell the master sergeant I was finished. When he looked up at me, he got this pissed off look. "Where the hell have you been all day?"

It took me a second but when it came out it was bad. "Cleaning your fucking latrine, master sergeant." Then my mind re-engaged and I said, "Excuse me." He looked like he was going to rip my head off. Then he stood up and said, in a calm low voice, "It fucking better be sparkling. Follow me."

I did. He opened the door, turned on the lights, and walked around with a, "What the fuck?" look on his face. Then he said, "Wait here. Do not move." I wondered what the hell I'd done wrong this time." He returned with the supply sergeant but walked right by me, out the door and back into the office area, where he hollered to captain something-or-other that he had to see this.

I stepped out of the way and the three of them went into the latrine. I could hear them clearly. The master sergeant asked the specialist if he had helped me. He said I had made a list. "I just drove him to pick it up, but that's it. I don't think he even went to lunch." The captain remarked, "I did not know we had a tile floor, did you?" The master sergeant answered, "No." Then the captain asked what my instructions had been. "Clean the latrine. That's all I said." The master sergeant then asked the specialist if anyone helped me. I was thinking I'd done something really bad.

Then they called me into the latrine. "How did you clean the ceiling and replace the lights that were out?" I said, "I got a ladder." The specialist confirmed it was on the list. The master sergeant told me to stay in the latrine and they all went into the office area. I was starting to get a little angry. He'd told me to clean it. If something was wrong, it was his fault, not mine.

The master sergeant came back and said, "Let's go grab a bite." We walked a half a block to the snack bar and went through the line. I had maybe a dollar and a half so I got a Coke. He said, you are not hungry? I said I was short on cash. He said, "My treat," so I got a sandwich. He asked why I didn't go to lunch and I said I didn't know where I was or how to get back to my mess hall. "You could have asked me." I said, "I was kind of wanting to get the job done before I ran out of steam or time."

He asked a bunch of questions. How old I was, how long I had been at Bragg, what I was doing here etc., etc. Then he shit on my parade. "You cannot start SF training until you are twenty-one years old on the day you graduate from your training. Didn't anyone tell you?" "No," I said.

I was absolutely crushed. I said, "I am starting to think I should have joined the Marines." He laughed, but I was not kidding. Then he asked, "How did you get assigned here?" So I told him how they wanted to make me a clerk and I had been promised I would get to go to jump school and get infantry training. I told him I wanted the opportunity to see if I had the ability to be one of the best. He asked me questions for quite a while.

When we got back, the place was locked up. He asked me to wait in his car while he made a phone call and would be right back. I was feeling kinda stupid and kinda like killing the recruiter. It was dark when he came out. He drove me back and we rode in silence. What could I do now? If I hung out here for my three years, I'd spend my military career cleaning offices.

Next morning, three of us were called out for transfer. I was back in the game: all I had to do was start lying about my age. We packed our bags and got on a three-quarter-ton truck to Smoke Bomb Hill. When the truck stopped, the driver only called two names, not mine.

"Where am I going?"

"Third group." He told me to report to the first sergeant. As he drove away, I was feeling very low.

I reported and signed in. I was escorted to a barracks, assigned a bunk and locker. I unpacked, made my bunk and sat on my foot locker, thinking this was horse shit. When the guys got back from wherever they had been they were friendly. Then we all went to evening mess.

THE DENTIST

The next morning, we fell out for formation. I had no idea what I was supposed to be doing. The first sergeant called us to attention, gave us the daily update, and then called out my name and told me to report to the orderly room following formation.

I did and he said, "You are assigned to third SFG's S-4, until further notice. Report there every morning until I say otherwise." I asked what that was and what I was doing. "I do not know what you will be doing. But I will have someone show you how to get there."

When the specialist 5 showed up it was the guy from the latrine clean-up. He took me to Master Sergeant Bragg and said, "Here he is, Sarge." He was on the phone and indicated I should sit, so I sat. I was feeling low. When he got off the phone, he said, "I could use a cup of coffee. Let's go to the snack bar." At the snack bar, he got a coffee and a doughnut. "You drink coffee?" I said I never had. He said, "Let me introduce you" and ordered another coffee and a doughnut. After I put sugar and cream in the coffee I kind of liked it.

We sat at a table in the corner and he started asking me questions. Where was I from? Where did I go to school? Why did I join the Army? I answered his questions and he would ask another. It was odd in hindsight but it was the first time in my life that anyone had asked me these kinds of questions. Eventually, he looked at his watch and said, "I want you to meet someone."

We walked between some buildings and into a back door. He asked me to wait there and returned with a captain. "This is Doc so and so," he said. "He is a dentist. When you finish here, come back to the office." I had never been to a dentist before. The captain asked me to follow him and I got in a chair and he started poking my teeth and saying something I did not understand to a female soldier sitting there. "Number xx remove, fill number xx." When he was done, he said, "Private Bailey, I want you in this office every day at 08:00. That is an order. Here is my office and home number. If for any reason someone tells you differently, you ask them to call these numbers and ask for me, do you understand?" I replied, "Yes, sir, I do." Then he said, "You have an unusual benefactor in Master Sergeant Bragg."

As I walked back to the office, I wondered what a benefactor was.

Master Sergeant Bragg asked if I had gotten any instructions. I told him and he said, "You are assigned to my section until further notice. When you get back to your barracks, if anyone tells you to go or do anything other than what the doc told you, tell them to call me and that I have given you orders to that effect. Do you understand?" I said I did. Then his voice changed and he said, "Go back and see if you can help Specialist 4 Johnson." Johnson and I cleaned the supply room, and after breakfast and morning formation I went to the dentist office.

The Dentist explained I had quite a few problem teeth and that he was going to remove the ones that were not salvageable and repair those that needed it. He asked me if I had a low or high tolerance to pain. I told him I didn't know, but it looked like we'd find out. He laughed, then said this was not going to be done in a day. "We may take a month or so, are you okay with that?" I said, "I have never been to a dentist, and I appreciate you working on my teeth, so whatever you want, I am ready when you are." He smiled. "Get in the chair."

First he gave me shots then he started pulling teeth. Every so often he would tell me to rinse and spit into a silver kidney-shaped bowl. It looked like it was mostly blood. It was not long when he said, "We are done with extractions. I am going to give you some pain pills to take when the pain starts to get bad, but before it gets really bad. They will make you sleepy. You are on bed rest for the next three days. You cannot eat anything you have to chew, liquids only, broth, ice cream, pudding. I want you to gargle with warm salt water starting tomorrow, three or four times a day. If something feels like it is not right, other than very sore and tender, or the bleeding does not slow or stop by tomorrow morning, then come back. Do you understand?" I nodded in the affirmative. I was starting to think that they must think I was really stupid with all this, "Do you understand," shit.

I stopped by the S-4, showed the sick slip to the specialist, and went back to the barracks. It was empty. I got my poncho out and covered the top half of my bunk and the pillow, went to the latrine and got a roll of toilet paper. Then I took a pill and went to sleep. I woke up maybe three times and took another pill and went back to sleep. When I woke up the barracks was empty. The guys must have been quiet as they fell out. I took a shower and simultaneously washed my poncho off. I was tempted to swish my mouth out but the doc had said to only gargle with salt water. I felt okay but I was getting hungry. I walked over to the mess and asked the cook if I could get some soup or ice cream. The mess was closed but he took pity on me and made me pudding. He sent a big metal bowl of it with me back to the barracks.

By the third morning, I was actually feeling pretty good when I went back to the dentist. He asked me how I was feeling, I told him I was fine. He explained he was going to drill the rotten parts out and put fillings or caps on. I said great, not really knowing what he was talking about. He did two or three teeth at a time. I went to see him for twelve to fourteen days at 08:00, then to work in the S-4 'til 17:00 or so.

I told Master Sergeant Bragg I appreciated him getting me in to see the dentist. "Don't thank me," he said. "Just prove you're worth it and we will be square."

GED

One day Master Sergeant Bragg asked, "Have you ever heard of a GED?" I told him I had not. He said, "Be here tomorrow at 07:00 and we'll see how smart you are." I was there when he pulled up and he told me to get in. We drove to the other side of the post and he introduced me to a lady who was the director of the education center and said he'd pick me up out front around 16:00.

She asked a bunch of questions: name, date of birth, home of record. I had to ask her what that was. She explained the address you put down when you joined the Army. Highest grade of school completed, high school attended and so on. Then she explained why I was there. "We will give you a series of tests to find areas you are weak in. Then we will sign you up for classes on those topics and finally we will schedule you to take the GED exam for your high school diploma."

I was really not sure what this was all about, but I figured Master Sergeant Bragg knew what he was doing. I took timed tests for the next two or three days. They sent them off to be graded and in about two weeks, Master Sergeant Bragg called me into the office and said, "I guess you're smarter than you look. Next Thursday you take the GED. Congratulations."

But what if I failed? I stewed on it for a few days and finally asked him what difference it made. I could see he was thinking. Then he said, "You, I think, are an exceptional young man. I believe you can become anything you want to be. I hope you become a great soldier. When you walked into my office I was not in a very good mood. I told you simply to clean the latrine. And then I actually forgot about you. When you walked back into the office after ten hours and said you cleaned it, I was pissed, because I thought you had been screwing off all day. When I saw the job you had done, I was, and still am, impressed. I said clean it and you, on your own initiative, did. I think you are worth giving a little help. So let's go get your high school diploma. It will come in handy no matter what you do. Trust me."

When we got to the education center there were three other guys there taking the test. We were all in one room with our tests, waiting for the start command. One subject after the other. I have no idea of how long it took, but each test subject was timed and it seemed to go by really fast.

When I walked out with the others, Master Sergeant Bragg was waiting. I was sure I had flunked it. I think he understood and asked only one question. "You want to get something to eat?" I was hungry.

The test, as I understood it, was mailed somewhere to be graded, then the results were mailed back.

A few weeks later, I was working in the back when he came in and said, "The results are back, let's go get them." I hesitated before asking, "Do you know if I passed or not?" He laughed and said, "It's your test, not mine. You are the only one they will give it to."

When we got there, two of the guys were waiting and the receptionist told us we would be called in and given the results one at a time. It seemed like a very long wait.

When I went in, she said, "Have a seat." I remained standing. "Could you just tell me if I passed or not?" She said, "I am afraid," my heart sank. "There is more to it than that." So I sat. She looked down and said, "It looks like you are in the blank blank percentile. And having joined from Denver, Colorado, that permits you to select which high school you would like to receive your diploma from."

Then she looked up at me. I had not heard much after the percentile. What the fuck was a percentile? She said, "Do you understand?" and I immediately said, "No, I do not understand. Did I pass or not?" She smiled and said, "Yes and you scored very high." I smiled then and said, "Thank you." I no longer cared what a percentile was. And getting a diploma from some high school in Colorado was not important to me. She said, "Your military records will have all this information. You are, as of a week ago, a high school graduate."

I walked out and said, "I passed." Then I turned back and asked, "Was it with honors?" She nodded.

I was grateful for what Master Sergeant Bragg had done for me, but now I wondered if there was a place for me in the Army.

As he drove me back to Supply, I asked him if there was any way I could get into an infantry unit since they said I was too young to be there. “I want to be a fighting soldier. What you do is important, I understand it takes a lot of guys to support the guy out front, but that is where I want to be.” He said, “I understand, but right now you need to learn to wait it out a little longer.” “Am I going to be stuck doing this for the next two plus years?” He said, “I think there are many ways to skin a cat.” I didn’t know what he meant.

A couple of weeks later, we formed as a group for our monthly payday formation, the entire 3rd SFG. We did this every month. They would have awards and decorations ceremonies, promotion ceremonies and the usual briefings on how great the unit was doing. Normally I would daydream through it. But then they started talking about the group's S-4 Master Sergeant Bragg.

The group commander was speaking. “I am sure most, if not all, of you are familiar with Master Sergeant Bragg in his role as our group’s supply NCO. But few of you know his history. I am not going to give it all to you. But I think that some explanation as to his new rank is in order. As a buck sergeant in Korea, he was assigned to a company that came under attack by what we call an overwhelming force. The company commander, first sergeant and all the platoon leaders were killed or seriously wounded. He took command as the ranking NCO and rallied his few remaining forces. He was wounded, losing the use of his left arm, but continued to command and direct both his unit's fire and the supporting artillery fire which stalled the enemy’s attack and prevented his unit from being overrun and destroyed. He received the Silver Star for his actions and was given a battlefield commission upon his return from the hospital. So I have the honor of presenting Lieutenant Colonel Bragg.”

Lieutenant Colonel Bragg walked forward in dress greens and everyone in the stands rose. The group commander called the group to attention and ordered us to pass in review. I never saw Master Sergeant Bragg again, but I do not think I ever marched as straight or as sharply as I did that day.

DOMINICAN REPUBLIC

I was eventually assigned to the 7th SFG and sent to the Dominican Republic.

But no one told me I was going to the DR. The first sergeant called me out of morning formation and told me to pack all my stuff. I packed my stuff and reported to the orderly room. The first sergeant said a Jeep would be by to take me to Pope Airfield. I was confused but simply said, "Yes, first sergeant."

It wasn't more than ten or fifteen minutes when my ride showed. I was putting my bags in the Jeep and the first sergeant came out and said, "Bailey." I turned and he smiled. I'd never seen him smile before. Then he stuck out his hand and said, "Good luck, soldier." I said thanks, and got in the Jeep.

I was wondering where the fuck I was going. I had not received any orders or any warning. Just, "Pack your stuff and get in the Jeep."

I got on a C-123, a twin engine version of the four-engine C-130. There was a Jeep and three pallets of boxes already secured. They were waiting for me. I strapped myself in. The doors closed and we taxied and took off.

I looked out the windows and suddenly we were over water, a lot of water. I was thinking that was one big-ass lake when I realized we were over an ocean. I did not look out again. It all seemed a little strange.

I went over the events of the day in my mind. I had been in the Army for about eight months. I always had orders when I transferred, and had never gone anywhere without someone telling me where and why. I was obviously leaving the country.

I do not remember how long the flight was. The crew chief told me to fasten my seatbelt and we started our descent. I figured at least I'd know where I was when we landed.

When the doors opened it was humid and hot. I walked off and waited to get my bags. A sergeant came up and asked, "You Bailey?" I said yes. He said, "I am in that Jeep, let me know when we get all your bags." I had my two duffle bags and an AWOL bag.

When we got far enough away to hear without shouting, he said, "I'm TAC. Thomas A. Carr, is my name but I answer to TAC." I told him my name and asked where we were. He said, "This is their airport." "No, I mean, where in the world are we?" "The Dominican Republic." He said it like I should know. "And the Dominican Republic is where exactly?" He looked at me like I was not playing with a full deck. "Didn't anyone tell you where you were assigned?"

"No, I was told to pack my bags and get in a Jeep this morning and now I am here." "That is fucked up," he said. "After you get signed in, I will explain where we are, why we are here and introduce you to a few of the guys."

When we got to the Special Forces Operations Base (SFOB) I reported to Sergeant Major Goodpasture. He said, "I heard good things about you, Bailey. Do not disappoint me." Then he told TAC, "He will bunk with you and Wadelick. Show him the ropes. And get that half-assed flash replaced."

The headquarters was in the old British consulate. The mission was to gather information on the Cuban arms shipments to rebel forces and stop them. I learned a lot there. And had my first combat experience. I was surprised at how calm I was and my lack of pre-op anxiety.

The Dominican Republic was like kindergarten or a first-grade introduction to combat. I was schooled on explosives both improvised and military issue C-4, composition B plastic etc. I was taught scuba diving, hand-to-hand combat, knife fighting and defensive position construction from hasty to permanent structure.

I unpacked and was taken around by TAC and Michael Wadelick (Mike). I was coming in after the initial wave and was kind of at a loss as to who the players were. As I recall there were five teams operating. TAC explained that the Cubans thought if they could get the populace aroused, which they had, they could take over. Their main problem was supplying their converts with weapons. Our focus was intercepting the shipments and killing those delivering them. They had Cuban agents and we had Cuban agents. Ours were mostly survivors from the Bay of Pigs fiasco that had joined the Army, then Special Forces, and now had infiltrated the rebel organization and were feeding us information on their operations.

Mike, TAC, Bob, and I were the security for SFOB. We pulled sentry shifts. And during the day I got demolition training, including recognizing and making improvised explosives and booby traps. We trained to rappel from both a copter and cliffs. There was SCUBA training and underwater demo, small arms maintenance, surveillance and assassination methods, overt and covert, ambush planning and, my favorite, hand-to-hand combat and the proper use of a knife.

I think I learned more in my time with these men than in the rest of my service.

One day I was asked to go on an ambush. One of our primary missions was to ambush weapons and ammo based on intel provided by Cuban agents. They would set up drop-off points at night and our Cuban agents would tell us the when and the where. I was allowed to participate in one. I am not sure why, but I think they wanted to see if I could actually bring it all together and do my job. I was one of the rear guards, meaning we protected the return route.

My role was small but critical. No one was allowed to see, hear, or be capable of identifying us. If anyone came into the kill zone, they were considered enemies and were to be eliminated.

I was invisible and positioned on our exit route, just inside a gate leading to the drop-off area. I carried a forty-five-caliber silenced grease gun and my K-Bar knife. I could hear the sound of someone walking on the trail toward me. As the sound got closer, I drew my K-Bar, which would be quieter than the grease gun. I crouched behind the gate in some scrub bush and as the sound got closer, I was sure it was only one.

Soon I could see the dim light from the candle lantern and then the man. He was older and walked with an uneven gait. As he passed the gate, I stepped out behind him, swung my left arm, grabbed his chin, pulled it up and back. Buckling his right knee with a kick, I plunged my knife into his neck and cut his throat, pulling the blade forward. The only sound was a gurgle and his falling to the ground.

I killed the light and waited. No one was following. I crouched next to him for maybe ten minutes. Then the muffled sounds of the silenced firearms opened up, lasting maybe three or four minutes. Then silence. I moved him off the trail and waited. I heard the movement of my team coming my way.

When they reached the crest, I could see their blue/green lights and turned mine on as I counted ten. I took up the rear position and followed them to the waiting trucks. The debriefing was quick. We had killed eight and brought out five or six cases of ammunition and four cases of rifles.

These men were trained and capable in their chosen profession. But there was a downside. I got to see who they had become. They had been through hell and seen too much, I guess. Most of them had no moral compass, or at least not one I could see. I still had values and believed in them. It took years for me to understand that those who answer the call to war are never really the same again.

Do not misunderstand me. I hold all those who, by choice or draft, stand ready to do what is needed, in high regard.

I don't really know if it was luck or intervention from Master Sergeant/Lieutenant Colonel Bragg that I was given that opportunity to learn directly from those men. But I am grateful for it. When I think about my rapid advancement in rank, and my survival (which I did not for many years), I know I would not have survived, nor would many of the men in my charge, but for this training. But youth is full of itself, and for a long time I just thought I was just a really good soldier.

When we redeployed back to Ft. Bragg, I stayed in the 7th Group.

There was a demonstration area named Gabriel, which is the place you see at the beginning of the Special Forces movie with John Wayne. Every so often demonstrations on our capabilities were scheduled. No one really enjoyed this duty. Standing at a station for multiple groups, giving the presentation and getting many of the same questions was boring. It did make us famous, though.

When President Kennedy came, the SF had been trying for years to get the wearing of the green beret approved by the Army Heraldry. He was so impressed by the demonstration that he told them to approve it—now. It made the national news. Prior to this, the SF was a kind of secret unit known only by the members of the military. About a year after that, Staff Sergeant Barry Sadler released his song "Ballad of the Green Beret."[33] We were now in the public eye.

The request for these demonstrations from senators, congressmen and other politicians caused a monthly need. And as with all popular things in the military, it kind of became a circus.

The last thing they added while I was there was the balloon tactical extraction. This was a high risk, and very dangerous, way of getting something, or someone, out of a very hot situation.

It was initially done with dummies. Then someone had the idea to have soldiers do it.

The process goes like this. The kite comes in a cylinder about four feet long and twenty inches in circumference. It weighs about thirty or forty pounds. You attach the rope to the item or person that is to be picked up. Then to the balloon. You fill the balloon when you contact the aircraft, usually a C-124 or C-130, and it rises about two or three hundred feet. The nose of the aircraft is equipped with arms and the balloon is caught between them and guided to a catch gear at the rear of the plane where it is attached to an electric winch and then hauled in. There are a lot of moving parts. A lot of things can, and do, go wrong.

After a few of these live extraction demonstrations, refusals to participate started to occur. So Commander Stillwell said he would do the next one. It did not go as planned. He was lucky he wasn't killed. He was jerked about ten feet off the ground, then the harness attachment failed and he fell, breaking a leg and hurting his back. After that the live demos were canceled.

Stillwell was eventually lost at sea. A plane he was on was lost somewhere over the Pacific Ocean enroute to Vietnam. No trace was found. It happened while I was in Nam in the 173rd. I was sorry about that. I think he was a good soldier and leader.

---

[33]Released January 1966

I reenlisted on a short, which means I had been in for over a year and a half and could reenlist for a unit of my choice. I was now about nineteen, and confident in my combat skills training.

I asked to be assigned to the 173rd Airborne Brigade in Vietnam. There was no way I was going to do details while a war was going on. I told no one. The day after I submitted my request, I was told to report to the group commander. I was not even sure where the group headquarters was, nor did I have a clue who the commander was.

It turned out the group commander was our SFOB[34] Commander in the Dominican Republic.

After he returned my salute, he asked me to have a seat. "Why are you leaving the Forces?" I explained I would not be eligible to go through S/F training until I turned twenty-one, and that detail was not what I joined to do.

He asked if it had anything to do with my tour in the DR. I thought about it for a minute and finally told him the other reason. I said, "Sir, I am not really comfortable with the idea of taking missions from secret non-military branches of the government or that we are not allowed to ask questions on the whys of our mission."

He said, "I understand why you feel that way. But have you thought about our country's need to protect our freedoms and our way of life from our enemies both in uniform and out of uniform?" I was thinking this was way above my pay grade. "Sir," I said, "the bottom line is, I want to see if I can actually be a soldier. I need to know if I can hack it. And your rules say I can't do that until I am twenty-one. But the guys fighting in Vietnam are eighteen." He smiled and said, "When you come back, if you want, call me, I'll get you a start date in less than twenty-four hours."

I got a ration of crap from a few of the guys, mostly asking if I'd lost my fucking mind. But I was in the Army and we were in a war. My job was inflicting pain on the enemy. I should have been doing my job, not executing the will of some unknown agency.

I was on my way home for two weeks' leave and then to Nam. I am not sure of the date. That is when I drove to Fort Dodge, Iowa, to pick up Melodye. We were married in Denver and I left for Nam about three weeks later.

[34]Special Forces Operating Base

VIETNAM

FIRST TOUR

The flight to Saigon was long. My first impression of Vietnam was that it was hot and humid. When they opened the doors, it was like opening an oven door. The heat took your breath away and was a stark contrast from the plane's air conditioning.

We boarded trucks and were driven to the replacement depot in Bien Hoa. We gave our orders to a specialist 4 and were then shown to a tent. A sergeant called our names and told us we would be called for a briefing in about an hour and to relax till they came to get us. We all went in, dropped our stuff on a cot, and laid down in the shade of the tent. I dozed off and I woke up to someone calling my name. I said, "Here," and a staff sergeant said, "Grab your stuff. I am here to pick you up."

As we drove, he gave me a brief introduction to the Charlie Company 1st Battalion of the 173rd Airborne Brigade. The drive was about an hour. He told me I would be processed at the brigade's headquarters. Then he would be back to pick me up in two hours or so.

When I got there, I found four other guys waiting to be processed in. Back then you hand carried all your records with you. This included pay, personnel and medical records. When you arrived, you hand delivered them to the finance office, personnel office and the medical unit that supported your unit. We were all waiting for our in-processing guide.

In this case it was the brigade sergeant major, affectionately referred to as Sergeant Major Blood Burns. He was about six feet tall and he spoke with a deep and commanding voice. He introduced himself and walked us through in processing. The last stop was finance.

I sent all my pay but twenty dollars a month to Melodye. They paid us in Military Script so US Dollars couldn't fall into the hands of the locals. U.S. dollars are easily used on the black market and can be sold for two or three times their face value in local currency.

Sergeant Major Burns told us to fill out a form with our stateside address on it and to give him twenty dollars, "For the year books." So I gave him twenty dollars. I was thinking what crazy son of a bitch would make a yearbook of combat operations. I mean really, who would ever say, want to see my war yearbook? I was in Germany about a year after leaving Nam and got the yearbooks, swear to god. We still have them.

I was assigned to the B Company, 1st Battalion 503rd Infantry. No one seemed interested in introducing themselves, which was fine with me. Processing was relatively easy.

When I took my medical records to the dispensary there was a specialist 5 there. I asked him for three hundred and sixty-four vitamins. He said, "A little pick me up, huh?"

"Something so I can keep track of my time and hopefully stay healthy." He gave me a bottle and told me it was exactly three hundred and sixty-five, the last one was for the trip home. I said thanks. He said good luck. Starting the next day, the first thing I did every morning was take a vitamin.

Our first days in country were mostly getting an orientation and adapting to heat and humidity. Then weapons orientation and boobytrap identification.

It all seemed to be poorly thought out and the instructors were not very knowledgeable about the topics they were teaching. But it was better than a learn-as-you-go program, I guess. I really did not pay attention to the names they came up with for our search and destroy operations. I was concerned about the maps and the intelligence briefings, both of which seemed to be lacking detail. Ditto for the operations missions which were pretty much the same for all of them. Seek out the enemy and destroy their ability to resist. I was not really comfortable with what I thought were poorly thought-out mission briefings. This was at the squad and platoon level.

I think my first major operation was named Junction City.[35] We went to the jump school in Saigon to go through a two-day refresher course before we jumped. When we got there, we were told that the 2nd Battalion was going to replace us. I and most of the other guys were disappointed. Turns out it was the only airborne assault conducted in Vietnam. I missed my only chance for a combat jump.

Vietnam was like the movies about WWII against Japan as far as terrain, rain, jungle, snakes and insects go. It was basically a hot, wet forested shithole—with guys shooting at you. We were, I think, the youngest fighting force the U.S. had fielded at that time.

We were all basically pissed off all the time, interrupted by moments of absolute rage. If ambushed, which happened four or five times, we responded with a kind of momentary insanity, as if the ambushers were the sole source of our misery and we just released all of our anger on

---

[35]Operation Junction City was an 82-day operation that began February 22, 1967. It was the first U.S. combat airborne operation since the Korean War and one of the largest operations of the war.

them. The Herd, as we called ourselves, operated mostly in war zone D, until we were moved to Dak To in the mountains to the north.

It was the longest and bloodiest fight of my tour. At one point they had two GP tents (the largest tents the Army used at that time) with body bags completely filling the space with room to walk between the dead. We were asked to identify any that had a white tag on the body bag if we could, as they did not have their ID tags on them when evacuated. It was difficult in some cases. Some had no recognizable faces. And some were missing major parts.

When we walked out after trying to help ID them, there were several large cattle trucks into which a detail was throwing the body bags with our fallen troopers inside, stacking them like bags of waste. Something snapped. I was pulling my Randall pistol out as I started toward them. When a major I had not seen grabbed my arm and said, "Sergeant, I am sorry for the way we must transport these men but time is short and they need to get home to their families as fast as we can get them there. Please understand, we mean no disrespect, we are soldiers doing the best we can under rather difficult conditions." I was ashamed of my loss of temper and told him so.

Graves registration has got to be the worst fucking job in the Army.

I think my luck at surviving and my previous training was the primary reason I was promoted so quickly. I went from an E-3 to E-6 in less than a year. At the time I never really thought about the promotions or why I was made a buck sergeant and squad leader so quickly. The squad accepted me as their leader.

We were busy conducting operations, training new arrivals and staying alive.

The men I ended up leading never gave me any trouble, even though some of them were there when I got there and all of them were older than I was. I was focused on the task at hand and doing it at a minimum cost in injury or death. I never stopped to consider how unusual a situation it was. I know now the time I spent with the 7th group prepared me better than whatever training the lieutenants and captains I was assigned to had gotten.

The 7th SFG training had given me the confidence to call horseshit when I saw it. I do not know if the soldiers in my squad, platoon or company liked me. It never crossed my mind until I was recently asked (I mean last week, June 23, 2018) if I thought they liked me and if I kept in touch with the guys I served with. I have never kept in touch with anyone from my family, or anyone from a unit I have served with, save one. And it was his persistence that really kept us in contact. His name is Rupert Finney. He will come up again later.

I was nineteen years old and an E-6, doing what I started out to do: I'd become an infantry paratrooper.

There were never any front lines nor safe areas; it was just, "Keep your head in the game and kill them before they kill you." We referred to the enemy as Sir Charles, out of a deserved respect for their skills and courage. I do not recall ever hating them. Which in hindsight seems odd to me. I guess that is what it is like on both sides. For reasons we never knew, we were told to go seek out and kill or capture those identified as the enemy.

Dak To was the first and last truly pitched battle of my tour. We were fighting a unit of similar size, equally equipped, who had had the time to fortify their position. It was a fucking nightmare. The brigade lost so many that we had to pull out and reconstitute in a place called Tuy Hoa on a beach somewhere. "Reconstitute" meant to get replacements for the dead and seriously wounded. We also had new guys that were not jump qualified, which put them at a severe disadvantage.

I pulled my squad together and asked how many jumps we'd made in Nam. Zero. How many were we likely to make? Zero I said. So the fact that some of our new fellow squad members were not paratroopers was meaningless. Our job was to teach them their job and welcome them as our brothers. Which they were. Were there any questions? There were none.

I was getting really short (meaning I would soon be DEROS, or Departing Enroute Return OverSeas). I was flown to our rear area to process. I was issued new uniforms and had to turn in my weapon and ammo.

That night around ten or eleven p.m. we woke up to explosions. I was wearing only my newly issued white underwear and I simply put on my boots and went out to see what was going on. There were explosions on the heliports and tanks were headed our way when the lead tank was blown to hell by a large shaped charge. The ones following were taking fire and retreating.

I saw a sapper[36] headed toward me and jumped into the Mess sump (a deep pit for food waste and trash) The sapper stopped and we locked eyes. I thought I was a dead man. He had his hands full of satchel charges.

---

[36]A member of an elite Communist unit known for surprise attacks.

He decided his assigned targets were of greater value. As he headed for the command tents, I got out of the sump, took off the white t-shirt and boxers and ran to the ammo storage site, bare-assed naked. I found the hand grenades, opened the crate and when I saw an enemy within range, I threw them and ducked.

I was watching when the Battalion Major startled me and asked what the fuck I was doing. I saw two more sappers, threw a grenade and we both ducked. We found their bodies the next day.

MANNHEIM, GERMANY

When I returned from my tour, I asked to go to Germany. Melodey had said she hoped we would get a chance to see Europe. I was assigned to A Company of the 3rd Squadron 8 Cavalry, 8th Infantry Division in Mannheim, Germany, at Coleman Barracks.

This was the oddest unit I served in during my Army career. And easily the most fun. An Army MOS (Military Occupational Skill) is an alphanumeric code that identifies your current job and past positions. I really never paid attention to what they had as my MOS. I probably should have, but I really did not care as long as they were letting me do what I wanted to.

My new MOS was 11B4P. This identified me as an airborne infantry scout. My rank was staff sergeant, E-6.

I was given command of Alpha Troops 3rd Platoon Infantry Squad. My Alpha fire team leader was Sgt. Jones; my Bravo fire team leader was Sgt. Spanglin. I had no combat vets in my squad. I was younger than most, if not all, of the squad. I was glad to be there and anxious to find out what they knew and what they needed to learn.

I had a couple of problems. My platoon sergeant was a fat ROAD (Retired On Active Duty) sergeant first class named Griffin and the other was the head shed (the company commander and the first sergeant). I came in Monday through Fridays to wake them and introduce them to my world and would pop in on weekends at odd hours to say hi, how's it going. My A-team leader Sgt. Jones was sharp but leery of me.

It seems the company commander, sergeant and squadron commander were just marking time. It was an odd squadron and an even odder troop. This was an Airborne (meaning we jumped out of planes) Armored (meaning we had tanks) Cavalry Troop with M-60 tanks and tracked APC's (armored personnel carriers) for the infantry. But we were also equipped with Jeeps with 106mm recoilless rifles for tankers and a three-quarter-ton truck for the infantry squads. We had alerts every other month or so and we had to wait for word on whether we would be airborne or ground forces, armored or wheeled transport. If by ground, we went armored. If by air, we went wheeled.

I had never heard of a unit with two completely different combat sets of equipment and mission readiness requirements. To further complicate matters, it was in the sixties and although I was aware of the race tensions in the States, I was not aware of it in the Army. I served with men who were black, white, brown and from Dothan, Alabama, New York City, Amarillo, Texas, Persia, Greece, Spain. We had arguments and even a couple of fights, but none of them were based on race or religion.

In Nam some of the units may have had race issues but I never saw it and it never happened with my squad. So this was a kind of "what the fuck, over" problem. I was very lucky to have Sgt. Jones and Sgt. Spanglin. They were both a little older than I was and they were both Black. I spent about two weeks watching and learning and planning with my two team leaders.

We had a particular problem: a private named Earthman. About six feet four inches tall and the self-anointed Black power leader in the troop. I conferred with Sgt. Jones and Spanglin and decided he had to go to preserve squad unity.

The following morning, I woke them up by tipping Earthman's bunk and dumping him on the floor. He jumped up and said, I am going to kill you. Jones asked Sgt. Spanglin to take the squad to formation. I said simply, you better get dressed if you plan to kill me. It was Earthman, Jones and me. I said let's finish our discussion in the basement. I indicated he should go first and when he was about halfway down, I grabbed the stair rails and kicked him down the stairs and grabbed the Matrix Handle I had placed earlier. As he started to rise, I struck him on the left chin then the right.

This really hurt and he started screaming threats and calling me a number of names. I asked Jones if he had gotten it (meaning recorded his threats). He said he had and I asked if he would call the MPs. And told Earthman he could be conscious or unconscious when they arrived. He chose consciousness. They arrived in short order and he was court-martialed and dishonorably discharged.[37]

---

[37]My father used to advise me that if I was ever part of a team, I had to get along with the group: if they listened to country music, I should listen to country; if they listened to rock and roll, I should listen to rock and roll—for my own safety. He also maintained that civilians are incapable of understanding what we ask of soldiers. When I first heard this story, I assumed it happened in the fog of war. To hear that it happened on a base in Germany, far away from any immediate

With Earthman gone, there was a sudden change in the squad. We became a team looking out for each other and an infantry fighting force.

Shortly after this, I was told I was going to the jumpmaster course. Jones and Spanglin kept the training and physical conditioning going while I was away. After the jumpmaster course, the platoon Sergeant sent me to the 7th Army NCO academy and lastly to the nuclear, biological and chemical warfare course at Baumholder.

When I returned, we had a new company commander, first sergeant and three new platoon sergeants. I asked Jones where everyone went. Seems they were all in a wife-swapping group and were putting pressure on the younger enlisted men with wives to participate and someone brought charges. I thought justice had been served.

I was told to report to the new first sergeant and did. He asked the same tired question about my age and I told him twenty. He said come with me. He knocked on the commander's door. We were told to enter. The new commander looked up and asked me a bunch of questions about my age and NCO status until finally the first sergeant said, "I think the captain wants to know how you became a SSG in such a short time." I said I was not sure why but the casualty rate and the long operations probably played a role. "If you need more than that, my battalion commander was LTC Sachs and my company commander's name was Captain Barnette. My last platoon leader is dead."

I was still standing at attention. He gave me what I assumed was an attempt at a glare. Then he said that is all. I did an about face and left. About a week later we had a change of command ceremony to welcome our new squadron commander and a new sergeant major.

I was the jumpmaster for a night jump and the new squadron commander, Lieutenant Colonel Davis, was a leg (meaning not a paratrooper). The new company commander had asked the new squadron commander if he would like to watch the jump with him. They must have just pulled names out of their ass when the wife-swapping scandal broke. But our new commander was obviously a suck-up, asking the new squadron commander to ride along.

When I was briefed by the first sergeant, also a leg, he said, make sure they get a good show. I said no problem. We were scheduled to jump from a C-141, a four-engine cargo jet, which was a little unusual. We normally jumped from propeller-powered C-130 or C-123. When I got the manifest, I advised the first sergeant that I felt we should

---

threat, makes me feel sick. I'm sure my father felt it was necessary at the time, but I question his judgment–even though my love for him is unconditional.

postpone the ride-along as the C-141 could be a little intimidating. He said something like, you fucking jumpers think you're special, you ain't. I doubt you will scare either of them. I said no problem.

There is a lot of coordination that goes into making a parachute jump at night. It takes a little more prep. And you need to identify those who are making their first night jump and position them early in the chalk for exit. The reason is once the first jumper exits, you really cannot react fast enough if a jumper has failed to hook up, which happens every once in a while. Add dim lighting conditions and you could have a dead trooper.

I had two assistant jump masters, RQ Woods and Barefoot. We went over the details and were comfortable. We loaded the trucks and departed for the military side of Frankfurt International Airport. It took about two or so hours to get the parachutes issued, on, and checked by one of the jumpmasters. We had about an hour and a half prior to scheduled take off. The jumpers were all resting back-to-back on the tarmac. The Air Force crew came out and gave us a thirty-minute heads-up. We got the troopers on their feet, the rear ramps were lowered, and we loaded. The last two on board were the squadron commander and the company commander. The seating has four rows, two outboard and two inboard. I requested the two inboards have one seat facing left and one facing right, placed directly in front of the jump doors. These two seats were for the two guests, so they would have an opportunity to see what the trooper sees and hear what the trooper hears as he exits.

We started our climb and everyone was in good spirits as I made my walk through. I had headphones on to hear the load master's warnings. I had asked prior to take off if they could open the doors at fifteen minutes out as opposed to ten. The doors were opened. The noise level went up. I am unable to fully describe the sound level, but it is necessary to scream to communicate, and you need to be standing next to whomever you're trying to talk to.

We were twenty-five minutes from drop. The interior lighting was dimmed to a low blue to allow the jumpers' eyes to acclimate to the dim conditions they jump into. The jet wash/wind flow is both loud and heavy. I looked at our guests, who both had wrapped their arms around their upright seats. The captain had closed his eyes. I was enjoying their discomfort.

What was about to happen was special. For jumpers, they have installed a special wind foil door that is put out at around ten minutes from drop. It has a Swiss-cheese kind of look but the holes are angled downward for two purposes: to push the trooper down and away from the jet stream so he does not spin or roll, and to force the trooper away from the fuselage. The last effect is the sound it produces; it is chilling, especially the first time you hear it.

Our visitors were still gripping the seat supports, but now their eyes were shut. They should have accepted my offer to brief them prior to the jump.

I gave the verbal (which was useless) and the hand-signal commands. Stand up, hook up, check equipment, sound off with equipment check. (In this case each trooper, starting with the last in line and moving forward, held up the thumbs-up signal. Then came the "stand in the door" command.

When the green light came, I gave a slap on the trooper's shoulder and the verbal command, "Go." They exited the aircraft very quickly. When both left and right chalk had exited, we pulled the static lines in and I looked at our guests. The LTC had his eyes open and a look of, "What the fuck was that." The captain still had his eyes closed. I signaled to Barefoot and R.Q. Woods to join me. We were standing in front of him when I signaled the crew members to close the doors. As soon as the spoilers were in and the noise level fell, he opened his eyes. We were all looking at him. He seemed to sense he had lost our respect.

We returned to Frankfurt. After we rolled out and came to a stop we three jumpmasters went directly to R.Q's car and drove back to Coleman Barracks in Mannheim. Then we waited (while playing poker) for the jumpers to get back.

When they arrived, we met them and had an informal debriefing. My last question was would all the first-time jet-jumpers please stand. They did. I asked, was my pre-brief of any help in preparing you for the jump? I was gratified by their response. Both Barefoot and R.Q. came up to me and said something like, kid your shit is rolled tight. I never understood exactly what that meant, but I did understand I had finally earned their respect. Which was nice.

We became tight and looked out for each other and helped when we could. I lost track of R.Q. Woods but heard Barefoot made command sergeant major and Sgt. Jones did the same.

About a month later I was told I would be going up for the E-7 promotion board. This was a surprise to me, I figured it would be three or four years at least. I prepped on my own and felt I was as prepared as I could be. I sat outside the interview room with about ten other selectees. They all were much older than I was. I was twenty at the time and I confess ignorant of a normal time frame to obtain different ranks. In fact, I never thought about it. I was just focused on being the best soldier I could be and taking care of and training to the best of my ability the men entrusted to my charge.

My name was called and I walked in, came to attention, saluted and said, "SSG Bailey reporting as ordered." There were seven or eight senior NCOs sitting at a long table. I was still holding my salute when the division sergeant major looked up, saluted, and his expression changed. "How fucking old are you?" he said.

I was accustomed to this. When I told him, he said in the same angry voice, "You ain't old enough to be a fucking corporal, let alone a sergeant first class. Get the fuck out of here. I did an about-face and started leaving. He said, "Don't you salute?"

I did not salute. I kept walking and said, "That would be a sign of respect," and left.

Flight School

I was pissed. I went directly to the S-1 (Personnel) and requested an application packet for officer candidate school and a packet for flight school.

I applied and had to give both packets to my company commander for his recommendation. He was not a fan, but from him it went to the squadron commander, who it turned out was, and then on to wherever. I was surprised when it took only about three weeks to get a response. I had been accepted for either OCS or Flight School. I had once again acted out of anger and frustration, but now I had to decide if I wanted this or was just blowing off steam. I decided I would go to flight school, primarily because most of the lieutenants I have dealt with were not very intelligent. Besides, I thought as a pilot I would never see them again. Plus, OCS started in six months and flight school required me to leave in about fifteen days.

I got my orders and was completely out processed in less than two days. All I had to do now was sign out. I got my tickets and we left Mannheim for Frankfurt early the next day. We landed in Texas, I think Dallas, and took a bus to Ft. Walters, Texas. I signed in the next morning and asked if I could sign out on leave to get my wife an apartment and pick up my car, which was arriving in a couple of days. The Major said you have two weeks before your class starts and they will let you know if anything changes.

I had my twenty-first birthday during the move or shortly after. The apartment we found was not great, but it was clean. We got settled in and picked up our car, a Lotus Elan. No one would insure it. I went and asked if anyone knew where I could find an insurance company that knew what a Lotus Elan was. One of the cadres said, come with me. We went into his office. He pulled out his wallet, got a card out and dialed a number. Then he told whoever answered, "I have a flight school cadet here that needs car insurance. Can you help him? Thanks." He handed me the phone. It was so easy and the rate was lower than I had been paying in Germany. I have been with USAA since and have no regrets.

My wife was settled and we just drove around and got our bearings. Once we got the TV working, the news was not good. It seems there had been a lot of fatal crashes and five cadets and three instructors had been killed in the past month. The investigators could find no apparent cause as yet. This upset my wife and I assured her I would be fine. But I really had no clue and was thinking this was not a good omen.

A few days later they announced they had found the cause. Seems the heaters were pumping carbon monoxide into the cockpit along with the heat. All the aircraft involved were the newer trainers, the TH-55. I was glad they at least knew why.

I was starting in a few days. This meant I would only get to see my wife on Thursday evening for about an hour. She had all the numbers for an emergency, but this would be our first separation since my Nam tour. And she was not a happy camper.

We had probably forty men to a flight, two men to a room and four flights in a company. I was in the 7th WOC (Warrant Officer Candidate Class). We were issued baseball caps and our WOC color was purple. At first it seemed a little corny. But I realized all the WOCs had shaved heads. We all ate at the same time in a very large mess hall. We marched to the mess but walked back on our own. I could see how someone right out of basic could get confused.

My roommate was Wayne Huddleston. A young man maybe nineteen or twenty. Friendly and bright.

Ground school academics lasted for about a month prior to meeting our P-1 flight instructors. Captain Hodgeboom was the flight commander. We were assigned in twos and threes to an instructor pilot. He would be our flight instructor for primary helicopter flight training. I made two friends in flight school: Theodore P. Smith and Robert Stern. Ted drove a Mustang. We met after he saw my wife drive off in the Lotus and asked, what is that? We got to talking and remained friends for over thirty years but lost contact when he retired from flying for the California Forest Service, fighting forest fires. Robert and I met in P-1 and were stationed at the Kon Tum Airfield in Viet Nam. He Flew AH-1s Cobras and I flew OH 58s, OH 6s and UH 1s.

VIETNAM, SECOND TOUR

My second tour was as different from my first as it could be. I had my own room with a fan. The officers' club was fifteen feet from my door. The showers and latrines were just like the barracks back in the States. We had daily missions and came back to base at mission complete to a hot shower, a hot meal, and quite often a live band or movie at the officers' club. It wasn't half bad.

ROCKET ATTACKS

My first rocket attack caught me in the shower. I knew we were supposed to evacuate the choppers and I ran to the ramp wearing flip-flops and a towel wrapped around my waist. I jumped in an OH 58, cranked it up, and took off with lights out. The rockets were still exploding on the airfield.

I was about five hundred feet in the air when it dawned on me that I had no helmet, which meant I had no radio, so no contact with the tower or any other aircraft. I remember thinking this was absolutely the dumbest thing I'd done so far. How was I going to avoid running into another aircraft and land the chopper?

We were supposed to evacuate to one of a couple of airfields in the area. I flew in small circles, maintaining my low altitude. Then the attack stopped. I flew right back and landed in the revetment I had departed from and shut her down. Then I walked back to my hooch. (I don't know why it is called a hooch.) I'd gotten away with it but I swore never to make that mistake again. And I never did.

The next morning at flight ops, my platoon leader said, "Mr. Bailey, what you did last night was brave, but extremely stupid, so I think we should just pretend it never happened." I never heard another word about it.

CLOSE CALL ONE

I had three memorable close calls as an aviator. In the first, I was flying a commander observing a helicopter assault on a Viet Cong Unit. The Cobras were firing rockets and mini gunfire on their position when a Scout OH 6 was hit and went down. The attack helicopters were unable to retrieve anyone and the remaining scouts were firing on the VC. I called, identified myself, and said I was bound to get the crew of the OH 6. The colonel in the back seat was talking on his PRC 25 and not monitoring my calls. I flew in and landed with the downed OH 6 between me and the VC. The pilot and crew chief got in, one in front and one in the back. We flew low and fast until we were far enough to climb to altitude. I flew the crew back, dropped them and went to refuel.

As they were refueling me, the colonel got out of the back seat and into the front seat. He plugged his headset in and said, "When I am in this chopper, who is in command, you or me?" I said, "You have the ability to tell me what and where you want to go. I have control as to if and how we do it. Does that answer the question, sir?" "Yes, chief, it does."

He got back in the back seat and we returned to the battle. It was winding down; the South Vietnam soldiers were closing in and calling for an aircraft cease fire. I dropped off the colonel and went back to Can Tho.

When I got to the operation center, I asked the ops officer if he had anything else I could do, as it was only around ten a.m. He smiled and said, "I think they want to talk with you," and pointed to my commander and platoon leader. "In my office," the battalion commander said. We walked in silence and when we got to his office I was asked to be seated and I did.

The commander asked what had I told the colonel and I repeated what I had said. He said, "Explain your answer." "As the pilot in command, I am given a mission, and once that mission is given it is my responsibility to complete it, as briefed, if possible, without destroying my aircraft or killing my crew or passengers. If, as an example, I arrive at a destination and it is being rocketed or under attack, I am the one who decides what to do, not the passenger. That is what I was taught is the meaning of PIC." They looked at each other, then back at me.

The commander said, "You are correct, but you put a colonel in danger without asking him." I said, "Sir, you were not there, and I submit we were both in danger. He asked to loiter so he could see and communicate with ground advisors on his radio. He was not monitoring mine. I saw an aircrew in deep shit and was in a position to get them out, and I did. If that was wrong, then I will turn in my wings now."

He said, "Let's just sit on this. You will not be assigned to fly him again. Okay?"

About four days went by before I was told to report to the commander. He was all smiles and said there had been a misunderstanding. He had been told what a great job I had done and how it had scared the shit out of the colonel. And the colonel had asked to have me fly him when possible.

CLOSE CALL TWO

The second was the scariest for me. It was during the monsoon season and I had called in the mission complete. The ops officer asked if I could pick up a guy and fly him to the POW camp off the coast. I picked him up and we flew to the POW camp. He got out and I called for clearance for take-off. They asked if I could fly an airman to Can Tho. I said I could but I needed to depart in the next twenty minutes to beat the thunderstorms, as I was not flying an Instrument Flight Rules-capable bird. "No problem," they responded.

They lied. I got off forty minutes later. Prior to take off I asked the tower to check with patty control on weather. They said it looked good and I departed toward the mainland. In less than twenty minutes I had no visibility, no radio contact and was regularly gaining and losing a thousand feet of altitude.

There was no going back. I dialed in patty control frequency and started trying to contact them. My doors were off and the rain became intense inside and outside the aircraft. I looked at the passenger, a young airman, and he smiled at me. This poor bastard had no idea the shit we were in and I was not going to tell him. I smiled back.

At twelve thousand feet, which was above my altitude limit, patty control finally answered. I gave him my situation and requested he vector me back to the mainland. He said, "I have you on radar. You are in a serious band of thunderstorms." I answered, "I am aware of that, I need you to keep me on course as my compass is fluctuating forty to fifty degrees." Which it was. He said he would stay with me.

The flight lasted about forty minutes before he said, "You should pop out in ten to fifteen miles." I asked him to keep me on course. He said to maintain altitude. I rogered and he said you are almost there and I popped out.

I could see the coast as I started my descent to the refueling dump. Suddenly it was as smooth as glass and we could see our landing zone. We were both soaking wet. I was alive and thanking God. My passenger was just smiling at me.

When we landed, I had to shut down the engine and secure the rotor blade prior to pumping gas. As I tied down the rotor blade, he said, "That was one wet ride." I tried to smile and said simply, "It certainly was," but I felt like grabbing him and telling this fucking idiot we'd almost died up there.

I dropped him off at the base of the tower and hovered back to parking. As I walked to our operations center, it occurred to me that the asshole at fault was me. I should have just shut her down and spent the night on the island. That was one of those a-ha moments when you actually learn something about yourself that can keep you, and maybe others, alive.

I'd accepted that what others had told me was correct. I was too mission-oriented and had a tendency to ignore or downplay the things that could go wrong. After that, I tried to adapt and look at ways to minimize risk and still reach the desired outcome.

And I relearned how to call bullshit when I saw it.

CLOSE CALL THREE

The next incident was an ash and trash mission. I was picking up two Marines to fly out to an outpost so they could coordinate naval gun targets for fire missions with the naval commander.

I was in a crappy mood. Bob Stern, my flight school classmate and friend, had been killed the previous day. Supposedly, the enemy ground fire had been too intense for any of his fire team to extract him or the copilot. And I had pissed off my commander by trying to fly out and recover his body.

The marine captains were not aware of my mood. As we flew over a canal with about seven to nine sanpan (these are wooden sail- and paddle-driven boats used to transport cargo on the vast canal waterways of southern South Vietnam), one of them said he'd sure like to get a closer look at those. And I took that literally. (At least, that is what I told my commander later).

I bottomed the collective and flew close enough to see their startled eyes and their clearing their decks for combat. I was flying fast and low beside the fourth sanpan when the tracers started flying by. I flew directly over one and pulled up to clear the trees. As soon as I thought we were clear I asked if it was close enough for the Marine to see what he'd needed to see. No answer.

I looked over at the one who wanted to get a closer look next to me in the front seat. As God is my witness, he had pulled himself into such a tiny ball that I could clearly see the full outline of the small armor-plated door that was maybe nineteen inches high and sixteen inches wide. He was in a fetal position, head buried in his knees. I still don't understand how he got so small. I reached over and pushed his shoulder.

He uncurled his body. His face looked white. I mean Scandinavian white, which is hard to do after a few months in Nam. He just stared at me for a while. As his color came back, I asked, "Are you okay?" He nodded. I asked again if it had been close enough. He nodded again, and no one spoke after that.

I descended to refuel. After I topped off, I walked over to them and said something like, "If I scared you, that was not my intent. But if you really wanted a closer look, it had to be right then." Which was true. I had no intention of scaring them or getting any of us hurt. For me, the request was doable, but it had to be done right away. I had to focus on getting down there and getting us out without any casualties.

Then I explained, "I've been here awhile. If I wasn't a little numb to risk, I'd be a latrine inspector." We got back in and headed for the outpost.

As I neared the area, I started calling to OP. I tried all frequencies, no response. I loitered for maybe twenty minutes and told them we were flying to Rock Jaw (it is not spelled this way), the area headquarters.

The same guy who'd wanted a closer look said, "There's another helicopter down there, let's just land next to him." It was then that I kind of bordered on insubordination. I said something like, "Are you really that new to Nam or just stupid?" He did not respond.

I changed my tone and said, "There is nothing in this country you can take for granted, nothing. Just because it was all good this morning doesn't mean it's all good now. If I can't get them on the radio, I am not landing."

We landed at HQ and I went in to talk with the S2 (This is an alphanumeric meaning "intelligence," which is not always a descriptor of those who work there.) I asked for the S2 officer in charge and eventually a major came out.

I told him what had happened and asked if he had any intel on their situation. In a very condescending manner, he said, "I communicated with them just a few minutes ago. They made all their SITREPS.[38] So, chief, the problem is on your end." I thanked him and walked back to the chopper. I said, "It must be my radio; their S2 just spoke to them. Let's go." When we neared the OP, I once again tried to contact them with no luck. So I explained we would be making a rather unorthodox landing, since I had to assume the landing zone was going to be hot.

As I flew over the landing area, I executed what is called a high overhead descent. At roughly two thousand feet, I dropped the collective and bottomed it, so we started sinking like a rock. About twenty or thirty seconds into the descent, we had four 51-caliber machine guns open up on us. White basketball-sized tracers were flying in front, behind, and beside us. I mean we could have reached out and touched them (but only

[38]Situation Reports

once). There are four you do not see between each tracer. We were now in their kill zone.

I continued my rapid descent and started my turn to downwind. At about a hundred fifty feet, I started trying to recover from the descent rate and as we were about four feet from impact we came to a hover. The captain was once again white and all scrunched up. We were so low I had to climb to clear the perimeter fence about nine feet high.

There were GIs running around inside the compound waving to us to come and land, which struck me as funny. What the fuck did they think I was going to do, go back up there? I was smiling to myself and realized the captain was looking at me with an open mouth. I cleared the fence and set down beside the UH 1 that the captain had suggested we land next to earlier. As I was touching down, I said, "Hey, captain, take a look at the Huey." It was riddled with holes. The rotor blades were shot to shit, along with the tail boom and fuselage.

I told them, "You have exactly twenty-five minutes until my departure. If you want to leave with me, set up another time for your coordination and get as much information as you can on their situation and most importantly the enemy's positions and estimated strength. I am pulling pitch with or without you."

They scrambled to confer with a lieutenant colonel, pilots and some others who were waiting for them beyond the rotors. I got on the radio to call patty control, gave them the situation and asked them to pass it on to my operations and to put it out on guard to warn other aircraft of fire in the area.

Then I switched to guard frequency (a frequency that is automatically monitored by all U.S. military aircraft in the country. My call was something like this: Mayday-Mayday, any fast movers in the vicinity of grid coordinates XX." I repeated my call several times. When someone called Bartender 19 asked how he could be of assistance, I responded with the current situation, the location and direction of the 50-calibers from the fire base, the fire bases' grid coordinates, and that they should assume SAM (surface to air anti-aircraft missiles) might be present. Now I had maybe seventeen minutes till pitch pull, assuming Bartender 19 arrived in the seven minutes he said he could.

Like a Swiss watch, they were on time. I gave them my recommended attack sequence: on their second strike I would pull pitch and they should not loiter because of the probable SAMs. They rogered and said they would give me a two-minute inbound call. I still could not see them.

I started opening my throttle from idle to go. They called inbound and the wing man followed. I saw the Marines running toward me. Bartender 19 called, "Two minutes inbound," and I was light on the skids. The two captains were climbing in when the first jet released his load. I screamed, "Buckle in, here we go."

As I started pulling pitch, the Marine in the front seat was locking the armored door. I hovered over the fence on the canal side and dumped the nose and pulled pitch as our speed grew. I popped up just above the trees and headed back to Rock Jaw.

As I flew, I went to a place I have been many times. It is certainly a place I know you are familiar with. I knew that the S2 major had to pay for what he had done. He knew damn well that it was his job to report when a base missed their SITREPS, but even if he hadn't reported it, he lied to me when I asked him about it. He put all of us in danger, neglected his duty and showed a callous lack of respect and remorse for the soldiers he was supposedly watching over.

The marine captain asked where we were going. I said, "Rock Jaw. I'm going to kill that major and then we will go back to Can Tho."

We landed. I got out and was walking toward the head shed, where the S2 major was supposed to be. There was a colonel, lieutenant colonel, and a sergeant major walking toward me at maybe forty yards. I kept moving a little to the right so we would not run into each other. They moved to their left so that we would. I transferred my .38 to my left hand so I could salute.

Usually, they return your salute almost immediately. They were not responding. I held my salute and again moved to the right. We were now maybe eight feet apart. What the fuck was wrong with these assholes? The colonel smiled at me and said, "You're Mr. Bailey, aren't you." I did not recognize him, but said, "Yes sir." We were now three feet apart.

"I would like you to meet Lieutenant Colonel XXXX." I dropped my salute. The sergeant major was now standing beside me. I transferred my pistol to my right hand and asked, "Do we know each other?" "No," he said. The sergeant major grabbed the pistol from my hand. I started to make a move but stopped myself. I asked, "Is there something wrong?" "No," the colonel said. I turned to the sergeant major and asked, "Why did you take my sidearm?" He responded, "Why did you have it in your hand?" I lied and said, "My holster got some dirt in it." The colonel said, "Mr. Bailey, your commander called me and he said you are a great pilot and a fine soldier and he asked me to tell you that we, meaning the chain of command, are aware of what has happened and that Major XXXX has been relieved and will be dealt with in accordance with the UCMJ. And that he would like you to bring the two Marines back now. Then you are mission complete."

I was just standing there looking at them for a few minutes. Neither I, nor they, said anything.

I asked the sergeant major If I could have my pistol back. He took the shells out and handed the gun back to me. I stood there and looked at each of them. Then I saluted the colonel and the lieutenant colonel returned it.

I admit that this is only my perspective and that it is possible I was wrong. But to this day I believe if I had been successful in killing that major the Army would have been better off and any court-martial board would have acquitted me of any wrongdoing.

I walked back to the bird. The two Marine captains were still buckled in. I said nothing. I just untied my rotors, climbed in and we flew in silence back to Can Tho. I landed, dropped them off and returned to parking.

As I hovered in, I noticed the battalion commander, my platoon leader and a chaplain standing by the revetments. I got out and walked toward my hooch. I was feeling very tired.

The commander asked, "Can we give you a lift?" "I'm fine," I said. The Commander said "Okay, then. I want you to get in the Jeep and take a ride with us." I got in. They drove to the officers' club, which I thought was weird.

We walked in and sat down at a table. The waitress brought us all beer and we ordered something to eat. I took a long drink of the beer, which was great. My battalion commander finally broke the ice.

"Joe," he said, "you have had a couple of really shitty weeks. I was talking with your platoon leader and he agreed with me. The maintenance officer is leaving tomorrow on R and R for two weeks and I want you to fill in while he is gone."

I took another long drink of my beer and said, "If that is what you want, it's fine with me. When do I start?" "Tomorrow at 07:00." I said, "I appreciate you taking me to dinner, but I have had an extremely long day and I am very, very, tired. May I leave now?" The Commander stood up and said, "Yes, and I am proud of your actions today representing the CAC (Corps Aviation Battalion)." "Thank you, sir, and good night."

MAINTENANCE DUTY

When I arrived at the flight line the non-commissioned officer in charge was just finishing his task assignments for the maintenance and crew chiefs. When they went off, he turned to me.

I had seen him before, but we had never spoken. I would come to the flight line, go to the aircraft assigned, do my preflight and if all was in order, take off for the mission. If I had an issue the crew would fix it, or I would be assigned another bird.

He said, "Good morning, Sir. I have been told you are the acting maintenance officer for the next two weeks or so. Welcome. I would like to explain two things. First, you are never to pick up a tool near one of my aircrafts. Second, once a day between three and five p.m., you will sign paperwork here in your office. Please do not mess with anything in the office other than to use the pen. This will only require fifteen to twenty minutes a day. I do not expect to see you other than between three and five p.m. Do you have any questions?"

I said, "I thought I might learn a little about fixing stuff." He said, "Sir. Do professional race car drivers fix stuff? Do airline pilots fix stuff? Do Air Force/Navy pilots fix stuff? No sir, they do not, and the reason is a little knowledge in a complex area will lead to killing someone. You relax and do something fun until it's time to sign the paperwork. Deal?" I said, "Deal. See you this afternoon."

I played tennis, slept late, had two appointments with what I think was a shrink. Second visit I asked if he thought I was a head case. He said, "No, but you do have a hard line on what is right and what is wrong. Your problem is that all the other people in the world don't know where that line is." Learning to play tennis and running around the airfield for two weeks was boring. I do not know if I was being punished or if the CO was just concerned about my headspace.

The shrink in the head shed must have cleared me because I was returned to flight status.

Taking Rob Home

I got called to the head shed a couple of days later. I reported and eventually the shrink called me in to talk. He had a somber look on his face.

He said, "I have a request. Rob's family has asked that you escort him home." I was wondering why. I had never met his family, other than his wife.

"I fully understand that he was your friend," the shrink went on, "but this job is probably the hardest thing I have ever done, and it will be the hardest thing you will ever do, if you accept the request."

He explained how the family would ask me what I thought they should do about a number of things and other members would blame me for their loss. "It is a heavy load to carry. I do not want your answer until tomorrow afternoon. Your missions for today are scrubbed."

It had been almost a week since Rob's body had been recovered. I had kind of moved on, which is what I do. I don't dwell on unpleasant things for long, I give what I can and bury it. This sounds a little nuts as I write it, but it is how I maintain my focus.

I really, really did not want to do this. I am not, as you know, a real people person and have no experience with consoling people in emotional pain. But I knew that Rob would want me to tell his wife and family he loved them. And that we found solace in each other's company.

The next day at noon I reported to the head shed. The shrink asked me what I had decided. I told him I would do it. He said, "steel yourself for this. You will represent the entire U.S. Army and I personally thank you for it."

I packed an AWOL bag and was gone in about an hour. When I got to Saigon, I gave a copy of my orders to the operations officer as directed. He read them and called someone's name to take me to the airfield now. He turned and said, "Sir, you will be leaving in about an hour. And thank you." I realized he was thanking me for escorting Rob home.

I was driven straight to the plane. I think it was a four-engine cargo jet. There were four other soldiers on board. The cargo was coffins. I do not remember how many coffins, but it seemed like a lot to me. We rode in silence. I do not think we even introduced ourselves. It was a long flight.

Dover Air Force Base is the recovery point for all those killed overseas. It is here that the dead are given their final preparation for burial and those assigned to escort duty receive preparation as well. It is a very somber place.

This is the one mission nobody wants. I think there were maybe twenty or thirty soldiers, sailors and airmen all together. We were given a small packet with information on our service members: Name, families' names, religion, military, family and personal history, education, marital status and children. We were briefed privately by one of the civilian staff.

Next, we were all measured for and issued a class-A uniform dress, socks, tie, complete with a name tag, awards and decorations, a dress belt and low quarters (shoes) that had been spit shined. We took a class on our duties and responsibilities as a military remains escort. It took about two days.

Rob was also being prepared.

I was asked to report to some colonel and did. He asked me to sit down and explained what I had already seen and knew. Robert was really fucked up. He had eaten the cyclic (his face had been slammed into it) and then been chewed on for the three days and nights we were not able to get to the crash site due to enemy anti-aircraft 51-calibers and heat-seeking missiles.

He told me he knew Rob and I were friends and asked if I knew the family. I said, "I have met his wife but that's it." He then blew smoke up my ass about how he had read my record and knew this would be hard. I was not in a particularly chatty mood and interrupted. "Can you just tell me what it is that is on your mind, sir?" He gave me a look. Then he said, "I think Robert would appreciate not being remembered by his family in the condition he is in. We've done all we can, but it isn't much. I have a photo if you would like to see why I think he would prefer a closed casket."

I declined the offer and told him I appreciated his kindness. I had already decided that if I could, I would convince them not to open the casket and to please remember that Rob was a proud man who would prefer to be remembered as he was in life, not in death.

The colonel stood up, put out his hand and said, "Thank you for doing this for him."

It took maybe three or four days to get everything ready. I met the hearse and we drove from Dover to New Jersey. The funeral director met us and put the casket on a stand in what I assume is like a viewing room. The director came over and told me Rob's father would be there in a few minutes and wanted to see his son. I told the director the coffin would stay closed until I talked with Rob's father. They told me that they'd comply with his father's wishes—whatever they were. I was very uncomfortable.

Mr. Stern walked in. Now he, Rob and I were alone in the room. I was standing beside Rob. His father looked at me as if I was the cause of his son's death and said, "Who the fuck are you, to not let me see my son?" Or words to that effect. I said, "I am here representing the U.S. Army. But more so, I am here representing my friend and your son. You can do as you wish, but first I will tell you Rob does not want you, or anyone that he cared for, to see him like this. I know because your son, my friend, told me on many occasions. He and I have seen more than our share of death. You, sir, are his father, and I mean no disrespect. Rob was a dear friend and you can put him on display if you want to, but I know that he would want people to remember him as he was in life, and not the thing in that box. I am here to fulfill my promise, to ask you to respect his desire to be remembered as he was by his family and friends."

I could see the anger flow out of his face as his shoulders slouched. The tension in his face and posture were gone and he looked like he was about to crumble. I stuck out my hand and said, "My name is Joe Bailey, sir. I have heard a lot about you and I wish we could have met under different circumstances."

He composed himself. As we shook hands, he looked me hard in the eyes. I looked back, not hard, just maintaining contact. It seemed awkward and long. Then he said, "Rob wrote about you. He thought of you as a good friend." I replied, "We were good friends."

He almost sat in the aisle, but I guided him to a bench and sat across the aisle. He took a minute to regroup. Then he said, "He will not be disturbed. You are staying with me." I tried to explain that I had a room and did not want to be a problem. He said it was settled.

We drove to his home, and he showed me my room and told me we were going to meet Rob's mother and sister for dinner. He would wake me in an hour or so prior to leaving.

I was tired but sleep did not come. When he knocked on the door, I was already dressed casually for pizza with Rob's mother and sister. His parents were divorced. His sister lived on her own. I was not looking forward to it.

We had been seated and were sipping our beer when they arrived. We stood and I was introduced to them. They were cold and there was no talking after the introductions, until the pizza came. It was Suzy, his sister, who asked why he needed a closed coffin. Her father had tried to tell her it was his decision, and they were getting a little hot and loud.

I said, "Miss Stern, if you will allow me. I would like to explain to you and your mother why your father has decided to keep it closed." She gave me a look I will never forget. Cold hate. I told them the same thing I had told his Father. I said, "I understand that I am not welcome, nor did I expect to be. I am here trying to keep my word to my friend. That is all I want to do. You are his family."

There were tears and then quiet. We had another drink. Then I went home with Mr. Stern.

The funeral went smoothly and I think Rob's wishes were carried out. At the funeral, his wife asked if I could come to her apartment that night for a little something to eat and talk. I was unfamiliar with New York, but Mr. Stern said he could drop me off if someone could bring me home. She said she could make sure I got back.

I dressed casually in slacks and a sports jacket. I was dropped off in front of an apartment building somewhere in New York City. As I neared the apartment, I heard music and talking. It sounded like a party.

I knocked, no answer. I knocked again, loudly. A man opened the door. As I walked in, I felt really out of place; all the men had beards and I saw no one else wearing a jacket. I stood by the door, looking for Rob's wife, and finally saw her coming from what turned out to be the kitchen. She saw me at the same time.

I started moving toward her and she handed a platter to another woman. She threw her arms around me and said very loudly, "This is Joe." The talking stopped. Everyone, and I mean everyone, some coming from other rooms, was looking at me. Then they all smiled and wanted to shake my hand and thank me for being such a good friend to Robert.

Rob's wife led me into the kitchen, where she asked everyone to give us a minute. She took my hands in hers and looked me in the eyes, smiled. "I may have not been completely honest with you when I asked you here. I want you to stand in for Rob. This is his son's circumcision."

I had no idea what this was going to require. I said, "If I can help, I will." She smiled and said, "Thank you for being Rob's friend. He wrote often about you and had great respect for you and thought of you as his best friend." I just tried to control my feelings and said thank you, he was a good man and we were friends.

She took me to meet the rabbi. He seemed very nice. He asked me how familiar I was with Jewish religion. He was taken aback when I admitted I knew nothing. He and Rob's wife went into the kitchen and came back in a few minutes. The rabbi came toward me with a furrowed brow and asked, "How are you at following orders?" I said, "Pretty good. Why?"

"In a few minutes we, meaning the chosen men of which you are one, will go into that room. We will perform a sacred ceremony for Jewish boys. The service will be in Yiddish. Just do what everyone else is doing. After the circumcision, I will fold the results in a cloth and present it to you. You are taking Robert's place for this. It is a very great honor and very unusual for a non-Jewish person to participate."

"What do I do with it?"

"You hold it in both hands and raise it about shoulder height for the others to see. Then we will sing a song while you hold it. Then the ceremony will conclude and you will present it to his mother and everyone will sing a song. Then the party will start."

"I will do my best not to screw this up, but if I do—please get me back on course." He nodded.

I remember the baby starting to cry but it lasted only a few minutes. No one corrected me so I assume I did it without major error. When I presented it, Mrs. Stern was crying. She gave me a hug and said something I did not understand and then the party was on.

I was emotionally and physically exhausted. I was driven to Mr. Stern's house around midnight. He was waiting for me. He said, "I want to apologize for the way I treated you and also for my daughter and wife." I said, "None required. I have no idea how my family would respond. I think it is a normal response to want someone to answer for a loss. I know Rob would have done it for me."

We had a drink and he drove me to the airport the next morning. We shook hands as both of us had tears in our eyes. I never saw him again. But Rob's sister Suzy became a friend and is still someone I love.

BACK TO NAM

I was back in Can Tho in seven days. It was odd, but I felt relieved to be back in an environment I had adapted to. I was not allowed to fly for about five days. It felt good to get back in the air.

I had about sixty days left on my tour, but I never had a short timer's attitude. Some guys got illnesses, eye trouble or any number of things when they had about ninety days left on their tour.

About two weeks prior to heading back to the States I got a mission to fly some lieutenant colonel to Saigon to go on R and R. I dropped him off then picked up a major looking for a ride to Can Tho and headed back.

There was a Chinook just in front of me. We were at about three thousand feet and just short of the Mekong River. The Chinook was maybe a half a mile ahead of us when I saw a SAM fired and hit the Chinook's right rear engine, blowing the rear rotor blades off. The Chinook's nose went up as it fell rear first toward the ground. Two guys had fallen out of the rear of the aircraft.

I went up on guard and sent the position and consequences. I did a high overhead to land beside the now-burning aircraft. We were getting some small arms fire from the tree line the SAM had been launched from.

I got a response from another aircraft in the area and asked if he could relay my position and report that we had SAMs in the vicinity. I told the major to get any bodies he could find and put them in the back of the aircraft.

Then some friendly South Vietnam Hueys showed up and started firing into the tree line with mini-guns. I was thankful for their presence and for the fire. The other aircraft brought back three bodies and said he could not get closer because of the heat.

I told him to get in, because the tanks would explode soon along with any ammo on board. I pulled pitch and stayed at about three feet off the ground until we were a mile or so from the crash site. Then I climbed and called my operations, told them what had happened, and asked them to send someone to help with the bodies. Fast movers, jets, were pounding the tree line.

When we got back, the ambulance was waiting, as was my platoon leader and the battalion commander. I shut down and explained what had happened and introduced the major who was covered in blood and other bodily fluids.

I asked him to relay what he had seen and done. As he started talking, I eased away and went to ops, completed my mission brief, and went to my hooch. I was standing in the shower when the captain walked up and said, "I was not finished talking to you, Mr. Bailey."

I continued bathing and responded that I had nothing further to say, sir. I was now standing there washing my nether regions and looking him eye to eye. He blushed and said, "When you are done, the battalion commander and I want to talk with you in his office, is that clear?" "Yes," I responded. I was not fond of either of them, and I am sure the feeling was mutual.

I reported to the commander. He asked, "Were you being fired on from the tree line during the body recovery?" "Yes, small arms, no 51-cals." "Did you return fire?" "No, the major was not armed and my .38 did not seem appropriate. But the eight or so Vietnamese Hueys sprayed the tree line with minigun fire and I did not see any fire after that."

"Why did you land knowing no one could survive?" "I landed to try and send something other than ashes home to the families. I was taught and believe that we do everything in our power to leave no one on the field of battle, dead or alive. Nothing in my flight training said, 'If you think they're dead, fuck it, leave 'em and run.'"

It took the captain a bit to recover. Then the battalion commander asked, "Do you think the major you had as a passenger would agree with you?" I asked, "Did you notice the CIB and jump wings on his uniform? He would agree." They looked at each other. A moment passed. Then the commander said, "That is all." I got up, saluted and went to dinner.

I learned years later from a pilot who was in the unit that they gave the major a Bronze Star for his actions, which was justly deserved. (I thought about this years later. Why did a major obey the orders of a W-1?) I do not know for sure, but I think he was thinking what I was thinking. Their loved ones deserved an opportunity to have something to bury and a place to go when they felt the need.

My tour was coming to an end. I flew my last mission the day prior to leaving. It was thankfully, uneventful.

AFTER VIETNAM[39]

I was assigned to a cavalry squadron at Fort Lewis, Washington.

I signed in and was told to report to the squadron commander, which I did. I knew when I walked into his office, he was a whacko or a wannabe. He had camo poncho liners covering his wall, a field phone next to a real phone, and the only furniture was a field desk. I was not impressed. When he looked up, his eyes were locked on my badges. Then he asked, "How did you get a CIB?"

This was the dumbest question I had ever been asked by another soldier. It occurred to me this guy is not playing with a full deck. I said, "By killing and or destroying the enemy's ability to resist as an infantry squad leader, sir." "How old are you?" I had thought I was past that. "Twenty-three." "How many flight hours do you have?" "Around a thousand or so." "Don't you keep a flight log?" he asked. "The ops officers record the hours you log in the aircraft logbook. I never saw a reason to keep a separate log."

He called the executive officer in and said, "This is Mr. Bailey. I want you to get him in the OH 58 IP (instructor pilot) course at Rucker." Then he looked at me and said, "You are my new Squadron instructor pilot. Okay with you?" I said, "If that is what you want."

The executive officer took me into his office, which looked like an office should. He got on the phone and gave them my stats (name, rank, and serial number). Then he said, your orders will be here tomorrow, your course starts next week. I saluted.

I drove home and told Melodye I would be leaving for Rucker for the instructor pilot course and would be gone for about three weeks.

When I got to Rucker and reported in, there were three other guys from Fort Lewis attending the same course from different units, a W-3 and two W-2s. I was the only W-1 in the class. We hung out together and we studied together. They did not seem to be bothered by my W-1 rank. We had a lot of fun and then I reported back to Fort Lewis.

After about a week I knew for sure the Calvary commander was not playing with a full deck.

---

[39]I asked my father to write this in the very brief window of our lives when I was old enough for him to tell me the truth and before his dementia set in. I remember hearing many stories from this part of his life that are not included here and I can't help but feel like he was racing against the deterioration of his own memories.

I started looking for a way out and found it with the Second Brigade aviation section. After I was told they would have me, I asked to see the Calvary commander. I said, "Sir, I do not think I am Cavalry material. I would like to transfer to the Second Infantry Brigade." As I'd suspected he would, he said, "If you ain't Cavalry, I do not want you."

SECOND BRIGADE OF THE 9TH INFANTRY DIVISION

I signed out of the Cav and into the Second Brigade of the 9th Infantry Division.

We were one W-3 maintenance officer, three pilots, W-2s, and me, the only W-1 instructor pilot in the division. We had two UH 1s and two OH 58s. I flew both. Our mission was to provide air support to the Second Brigade.

As an instructor pilot I had to take a check ride with one of the Standardization Sips.

When I showed up at the appointed time, a W-3 asked me what I needed. I said, "An IP check ride so I can do my job." He said, "You have to go to Fort Rucker and pass the training course first." I replied, "Been there, done that," and gave him a copy of my course completion. He handed me off to a W-4 for my check ride.

During the preflight the W-4 asked, "How many lubrication points on the skids?" I said, "None." He asked are you sure? I responded yes. He said, "You know if there are, I will fail you." I said, "If there are you should. Shall we see who is right?" He smiled and said, "No, let's fly." He ran me through the maneuvers and said, "You're pretty good. Can you fly between those trees?" I said, "I can, but it is not in the check ride maneuvers." He said, "Show me." I did. He said, "I did not think you would do it. Let's go home, you pass."

After that I was tasked with giving check rides to all the units as requested and to train the pilots in my unit.

It was awkward when I showed up to give a W-3 or W-4 a check ride; it was even worse when I grounded them for lack of technical knowledge or flight deficiencies. I became known as a hard ass. My own flight commander, Jack Caulder, would take me aside and say, "Joe, you remember me telling you about catching more flies with honey than vinegar?" It was always about me hurting someone's feelings or my lack of respect for a higher rank.

So I just stopped talking to anyone. I just flew from point A to B. Then he told me I was not being cordial to the passengers. I explained I was not driving a limo or taxi and they were not paying or tipping me for a service. I was just doing my job.

Jack was patient with me. But I did not change, and even though he was getting complaints from the Cat B officers about me, I was getting great officer evaluation reports.

Cat Bs were officers that were pilots but were assigned to other duties not including flying. So to get their flight pay, they had to pass a check ride every six months. If the ride was not good, they had to go through training to get back up to speed. When that happened, they caught hell from whoever they were working for and some got bad evaluations.

At the time, commissioned officers got more flight pay than warrant officers, who flew all the time. I felt that if you wanted the pay, you should maintain the skill or give up your wings. So if they did not have their shit together, they got a no-go from me on the check ride.

Jack A. Walker

We got a new brigade commander named Jack A. Walker, who was a very young man to be a full colonel. About three months after he arrived, Jack had the tower call me via radio and told me to return to base.

When I arrived, he asked if I had a set of fatigues here. He seemed very antsy. I didn't have my fatigues. He said, "We need to get to brigade headquarters. It's a mandatory formation." I told him, "No one told me." "I know," he said, "no one told me either." So we all got in our cars and drove to brigade HQ.

When we got there, the S-1 A captain, who was a nice guy, ran up and grabbed me and said, "We need to go, come on." I looked at Jack and he shrugged his shoulders. I went with the captain and he left me standing in a formation of a group of two or three guys in front of the battalion formations facing the brigade staff. I felt like an absolute ass and had no idea what I was supposed to do.

The brigade commander and staff marched out to the field. The commander called the brigade to attention. They had a microphone setup and he walked forward and gave a spiel about recognizing soldiers who have gone above and beyond, etc., etc. Then he gave the command for awardees to center march and the group I was in started forward. I had to catch up to them and felt like an ass again.

I was positioned last on the right. The commander stopped in front of the first soldier and the S-1 at the mic read the orders for the award. I had no idea what I was doing in this formation and I looked like I'd just rolled out of a duffle bag. Then Colonel Walker stepped in front of me and the S-1 started reading, "Attention to orders WO 1 Joseph E. Bailey blah blah." I was trying to figure out what I was getting an award for. Had I been listening, it would have given me a clue. Colonel Walker said, "Mr. Bailey, I am honored to present this award." Now I was back to paying attention. He stuck out his hand, I shook it and said, "Thank you, sir." The ceremony concluded and I walked over to Jack and said, "Someone should have told me I was getting an award and to wear fatigues." I was a little pissy. He said, "Joe, honest to God, I got the call a minute prior to calling base ops and telling you to return to base."

I walked to my car and drove back and completed the check ride I had started. Then I went home.

The next day everyone seemed to be avoiding eye contact. I finally said, "Am I dressed funny or what?" Jack said, "We just don't want to piss you off." I said, "Thanks, but that was yesterday. I apologize for being a little pissy, but I felt like an ass out there in a flight suit."

Colonel Walker told me to report to him, and when I did he said, "Mr. Bailey, I have a problem and I think you can fix it for me. I have a battalion commander who is afraid of flying and I need him to get a bird's eye view of the Yakima firing center prior to our war games next month.

I would like you to convince him that you will not scare the shit out of him or kill him."

I asked why he thought I would do either. He said he thinks all pilots are showoffs. I agreed to try, and he said, "You have an appointment with him in thirty minutes."

As far as I was concerned, if this guy didn't want to fly, I couldn't make him. But I reported and for the next hour relived the commander's experiences with pilots and the carnage they caused. Then he asked, "What makes you different?"

I said, "Number one, I am a pilot because a division sergeant major pissed me off at an E-7 promotion board. Otherwise, I would still be in an airborne infantry unit, not providing taxi service via helicopter. I do not do what I do because I think it is cool or exciting. I do it because I am trained to do it. I find excitement on the nights my wife is feeling sexy. The rest of life is pretty much a job. I am not here to con you into flying with me. I am here to tell you, if you decide you need to fly, that I will be the guy you can trust to do everything I can to make your flight boring but productive. It is your call, sir." Then I stood up, saluted and left. I didn't give a shit if he wanted to fly.

When I got back to my operations, Jack said, "The brigade commander wants to see you now." I drove back to his office, thinking I had pissed off the battalion commander. I reported to Colonel Walker and he asked what I'd said to him. I said basically that I was no hot-shot pilot, but it was his choice. He smiled and said, "He wants a trial run tomorrow to Yakima and back around 08:00. Thanks, Joe, I knew you could do it." I walked out of that office thinking I was right not to become an officer. They are wound way too tight.

I picked up the commander and we flew to the Yakima firing center. I called and got clearance. The ranges were all clear (not in use). I had the commander wearing a flight helmet so he could monitor my radio and wouldn't interrupt when I was listening or sending.

We refueled and I showed him the entire range and showed him the drop-ff pilots like to use to scare passengers. I told him if he got another pilot and he headed this way, to just say, "Been there done that, asshole, so let's try and be professional." He asked if I ever flew anyone over the ridge. I said, "I fly everyone out here the same way I did with you. So they will not be spooked by some asshole who happens to be a pilot." We flew two fuel loads and headed back across the Cascades to Fort Lewis.

I said there might be times the weather forced us to fly low through the passes. So I took him down and flew through White pass. I explained if the weather was bad, we could come to a hover, turn around and go back, or find a spot and land till the weather breaks. I did all of those. "So if someone is giving you the willies, tell him to turn back or land. He may know how to fly, but you outrank him. Don't ever think you are at his mercy. And if any of our aviators ever scare you, let me know. I will return the scare and we will never leave the ground." We landed back in front of the brigade headquarters. And I said, "Thanks for trusting me." He did not respond, just unbuckled and got out.

I refueled and parked the bird. When I walked in, Jack said, "The colonel wants to see you." I had just about had it. "Jack, are you not my immediate supervisor and as such, is it not your job to run interference with the head shed so I can do my job?" He glared and said, "I am also invited."

We reported and the brigade S-3 came in and set up a bunch of maps and briefed us on our mission over the next few months. First, we would be the aggressors against our own division, then against the 101st airborne division with a two-week break between them. We would stay in Yakima for the duration and in the field.

We stood to leave and the colonel asked me if I could stay a minute, like I had a choice.

He asked if I would be willing to act as his pilot for the duration of the exercises. I said, "Sir, I am the section SIP and It is my responsibility to keep an eye on our aviators for signs of stress or fatigue and to make sure they are up to the tasks they are given. If I can stay with them and not in the HQ, I do not have a problem with it, but when I am mission complete, I want to stay with our section."

He smiled. He did that a lot. "Deal," he said. I added that we needed to have a radio telephone operator crew so we didn't have sleepy or tired crew chiefs or pilots. He said, "No problem."

The brigade whipped both the 9th Infantry Division and the 101st Airborne Division. A day or so after the maneuvers were over, we were still in the field when I was told to fly over to Yakima municipal airfield and pick up the chief of staff of the U.S. Army and fly him to the field HQ of our brigade. I did, and when we landed Colonel Walker met us and asked me to land somewhere private. I had the entire firing center to choose from, but I thought this was important, so I landed on a high flat-top peak. I shut down and the Army chief of staff and Colonel Walker walked off maybe fifty yards and had a nice long conversation. We napped and the chief of staff's aide kept watch. I guess it lasted an hour or so. When they returned, I dropped Colonel Walker off and took the chief of staff back to Yakima Airport.

(This is a rather lengthy segment, but I think it is important.)

The next morning when I landed and went into the headquarters, Colonel Walker said, "Let's go for a walk." When we were alone, he said, "Joe, do you know what the chief of staff told me?" "No, sir." "He told me I have been selected to be promoted to brigadier general and that I should not be surprised that I will not be able to make the changes I think the Army needs. He said, 'I am the chief of staff and I can't even change the head gear.'" I congratulated Colonel Walker and he said, "Thanks, Joe." That was the first time he called me by name.

A few months later I was told by the S-1 to show up at Colonel Walker's house on post with my wife. I asked him to call the colonel and tell him I had no desire to go. He called me back in about fifteen minutes and said, "Sorry, Colonel Walker specifically asked you to be invited." I asked if that meant I could say thanks but no thanks. He said, "No, it means get your dress blues ready and be there." Melodye for some reason was happy to go, but I told her we should pay our respects and get outta there.

When we got there, the street was full of cars, so I parked in the alley behind his house and we walked to the front and rang the bell. Mrs. Walker opened it and we went through the receiving line. Jack and his wife Kathy were a few couples ahead of us. I introduced Melodye and thanked them for inviting us. Then I found Jack and said, "I am very uncomfortable. How long do I have to stay here?" He said, "Long enough to eat something." Melodye and I got in line. They had people to serve you and they put so much shit on our plates it was embarrassing.

After we finished the line, there was no place to sit. We were standing by the kitchen and the back door was open. I said, "Let's go eat in the car." So we did. We were sitting there happy and actually joking while we ate when Colonel Walker knocked on our window and startled us.

I rolled down the window and he asked, "What are you doing?" I said, "Sir, there were no seats and we did not want to try and eat standing because it is very crowded. We are coming back as soon as we finish, sir." He said, "Nonsense," and opened Melodye's door. "I will find you a seat." I was fucked. We followed him carrying a plate in one hand and utensils in the other. He walked and he said loudly, "I need two seats for Mr. and Mrs. Bailey, please." Everyone stood up. I was so embarrassed I wanted to just melt away. Jack had a shit-eating grin on his face and Kathy looked like she just wanted to run over and rescue us.

Then it started. Colonel Walker started telling everyone there what a key role I had played in the brigade's victories over both the Corps and the 101st. How I had done what he thought could not be done, reminding everyone he was also a pilot. I was trying very hard not to show my discomfort. (Jack later said I looked like someone was holding a branding iron to my ass) Then, as God is my witness, he asked one of the Battalion commanders to get Melodye and me a glass of wine.

Finally, I saw his wife coming to our rescue. She put her hand on his shoulder and the communication was clear. He apologized to everyone for going on and then ended by saying, "Mr. Bailey is a soldier's soldier." Then everyone seemed to forget about us, except Jack and Kathy. They came over and Kathy started talking with Melodye. Jack just put his hand on my shoulder and said, "He meant no harm." I just wanted to get outta there.

We left shortly after. On the drive home Melodye said, "He thinks very highly of you." I said, "That was embarrassing." She said, "I think he meant what he said." I replied, "I do not care. When I got to work on Monday everyone is going to think I'm a suck-up. And I am not. So fuck them all."

The next time I flew Colonel Walker to Yakima for some meeting, we must have been about halfway there when he said, "Joe, something happened over Christmas that is bothering me. Can I talk to you about it?" I said, "If you want to, but I doubt I can do much good." He said, "I married into a very wealthy California family. We were visiting my wife's family and they traditionally host a big party around Christmas for friends and family. I was there once before we were married, and after dinner the men all went into the den for drinks. I was not asked then, but this time, I was. I would guess there were maybe thirty or so there. They started talking politics and his father-in-law said 'we have decided we will put Ronald Reagan in the White House.' I was a little taken aback. I have heard about the money men, but I really never thought they actually existed. Does that bother you?"

I said, "Are you talking about the actor?" "Yes," he said. "That's their candidate for the next election. An actor." I said, "We have had presidents from all kinds of backgrounds. I don't think we need to worry. And how bad can they screw up in four years?" When Reagan won the nomination that year I was taken aback, but when you think about it, it seems logical that people with money have the power of advertising, which seems to be what elections are swayed by.

Colonel Walker was promoted and left for a finishing school for new generals. Five months later Jack called me in from a check ride and said, "The Corps adjutant has asked that you report to his office now." This was not normal, a W-2 (I'd been promoted) being asked to report to a Corps adjutant. I drove to the Corps headquarters and asked directions to the adjutant's office, still dressed in my flight suit.

I finally found his office. "Oh yes, Mr. Bailey," the receptionist said, "have a seat and I will tell him you are here." There were four or five full colonels in class A uniforms waiting to see the adjutant and I asked if maybe I should come back tomorrow more properly dressed. She said, "Oh no, he wants to see you as soon as possible." There were chairs in the area where the colonels were seated but I just sat in the chair next to her desk. The colonels were eyeballing me like I had pissed on their boots. I did not recognize any of them but hoped I was not here because I had insulted one of them.

The door opened and one of the colonels stood up and the receptionist said, "Sorry, sir. He wants to see Mr. Bailey first." He started to speak but the door opened and out walked Jack A. Walker wearing a star. Smiling and holding out his hand, saying, "Joe, it is great to see you. Come on in." The room was quiet. Two of the colonels' mouths were hanging open.

After pleasantries, he said, "Joe, I would like you to be my aide. I need someone I can count on and I am sure I can get Congress to give you a congressional appointment to lieutenant, having looked at your records. What do you say?"

I thought about it for maybe thirty seconds. Then I said, "Sir, I have never been prouder than I am here and now. You will never understand how much you asking means to me. But I was offered a commission before." He said, "I know, I have read all your files." "You are honest, educated, sophisticated and so many other things I am not. Things are simple for me. There is right and wrong. I have no middle ground. You will need someone far more polished than I could ever be. But I honestly want to thank you for asking me. I have one other thing to ask. Please allow me to fail or succeed on my own merits. I have heard of guardian angels in the Army, and I don't want one. I want to know that however far I get in my career, it was on my own merit and not because I had help from someone along the way. But, sir, if you ever need me as a soldier or friend, I will be there. I hope you understand."[40]

He came around the desk and grabbed my hand. "I understand and I am glad I know you. Good luck, Mr. Bailey."

I never saw him again. But when I was selected for the warrant officer advanced course and showed up to sign in at the mailroom, I had a letter waiting. It had a red flag with two stars on it. When I opened it, it read, "Congratulations, I had nothing to do with it. But I am proud of you. Your friend, Jack A. Walker."

SIP AS A W-2

I returned to Germany as a W-2 IP (Instructor Pilot). Upon my arrival, I was taken aback by the lack of military bearing or discipline in the unit as a whole. I arrived with Al Chapman and Rupert Finney, two instrument flight examiners. We hit it off, but the guys already there were not my kind of guys.

I went to the 7th Corp's standardization section for my in-country check ride. It took about three days to brush up on International Civil Aeronautics Organization flight rules. The rest was flying and I was gone for one day. When I got back, the commander, a major (I forget his name) called me into his office and asked me if I would be willing to become the Corps OH 58 standardization officer.

I was a W-2. There were two IP, W-3s senior to me in rank. He said neither one would do it. I said, "This is still the Army. If you tell them to do it, they have to do it or they get court-martialed. Isn't that still how it

---

[40]I'm not sure what my father is trying to communicate here. He was always mistrustful of management, politicians, and anyone interested in power, but often took leadership positions with his unit. He valued leadership but hated politicking. He loved the clarity of a kill or be killed situation.

works sir?" He said, "I don't want to force them if they won't do a good job." I said, "Call Mr. Chapman and Mr. Finney in and let's talk."

When we were all in the room I said, "There is an SIP position vacant. Do any of you want the job?" I asked each in turn and they all said no. I said, "If I take this position, I will run standardization for this unit and I will be your senior by position, not rank, but position has the same UCMJ weight as rank. Are we clear?" They nodded in the affirmative. "If I say this is what we are going to do, then you do it, or I will bring insubordination charges against you. Are we all in agreement?" They said they were. I said, "Okay."

I flew up the next day. I took the written exam and the oral, then the flight portion. I was now the Corp's SIP, as a W-2. It was fun. I went all over the Corps giving check rides and we set standards for new guy orientations and prepared them for their in-country check rides.

I found out after about five months that they were swapping wives and had no moral compass, I guess it's a thing a lot of people did but I wanted nothing to do with it. I was pissed. I went to the battalion commander and asked him if he was aware of the goings on and he said it was not his job to teach Sunday school.

I was looking for a new home when we had a change in company commanders. It turned around pretty quick.

I deployed with the scout platoon for field exercises. We were doing a division winter field exercise. We were all in a large tent and I was using flashcards to quiz the pilots and crew chiefs on target identifications. The section commander, a first lieutenant, came running in and said, "The Corps standardization and their colonel is headed our way." He came in.

I introduced him and said, "If you find room for improvement in the training program, please let me know, sir." Then I left. The after-action report had a comment from him saying he was very impressed with the scout platoon's knowledge and commended the platoon leader on his training program.

I made W-3 and was selected for the senior course. Which is like a six-month course for the top-rated W-3s. Then back to Germany then back to Fort Lewis where I made W-4.

This time I could not escape being assigned to the Cav squadron. Once again, the unit was commanded by a lieutenant colonel with a really bad inferiority complex. We did not hit it off and the friction affected the unit's morale. We had two suicides.

Then we went to gunnery practice at the Yakima training area. (This is a graded exercise that gun pilots have usually on an annual basis. And there is a minimum score requirement.) I noticed every crew was qualifying on their first range run. So being me, I looked into it and found that this was an order from the squadron commander that all crews would qualify their first-time down range.

I was pissed, but thought, "There is no war, and maybe the Army is only issuing enough ammo for one run." After we got back to Fort Lewis there were about thirty Conex containers in the barracks area. When I asked about it nobody had any idea why they were there or what they contained. Being the curious guy I am, I found out.

They were full of live rockets and mini-gun ammunition. I went nuts. First, I went to my company commander, who was an absolute kiss-up pussy. Then the squadron commander, who told me it was none of my business. I then went to the brigade commander and he said the same thing.

I am not comfortable with unnecessary risks to soldiers. Every day they are exposed to life threatening tasks. But for a commander to put live high explosive ammunition next to their sleeping quarters and hide the danger from them was, to me, criminal. So I got on the phone and called my branch and luckily the man who answered had gone to the advanced course with me.

After I explained my situation, I told him, "I am going to the Corps commander and the local newspaper. How fast can you cut orders to send me somewhere?" He said, "I can cut orders sending you to the safety officers' course and then the Apache course. You will have them in two to three days." I thanked him and made an appointment with the corps executive officer for five days later.

Seven days after the shit hit the fan, I was packed and enroute to Fort Rucker for the Apache Helicopter 64 (AH 64) training, and the safety officers course. When I was back in Germany, one of the pilots sent me a local newspaper article about the breach of trust of the chain of command in placing so many soldiers in harm's way. And that both the brigade commander colonel and the squadron commander were forced out of the Army.

Desert Shield Desert Storm

I was selected to be the brigade's aviation safety officer. Then came Desert Shield Desert Storm. I was a staff officer by then. The only thing of note was my confrontation with the two battalion commanders on the aircraft parking alignment.

The space was tight and my plan required one AH 64 to park in the line of the other unit's area to minimize the risk of an accidental weapons discharge blowing our aircraft to hell. I explained to the brigade commander the whys and ended my briefing by stating that, "Eventually someone will accidentally fire a rocket or hellfire missile."

Then one of the commanders said, "There is no way that could happen. If anyone accidentally fires a hellfire missile, they should get a medal." I said, "In my opinion it is an unnecessary risk to park them nose to tail." The brigade commander agreed with me.

The next day someone accidentally fired a hellfire missile.

Its trajectory took it between two parked Air Force cargo planes, in front of an Air Force plane just landing, and into the Air Force's ammunition dump. On impact, the ordnance started to cook off and continued for about nine hours. No fatalities but a total loss of bombs and ammunition for the Air Force.

That night when I gave my briefing, I started by asking, "Are we here to offer a medal to the ground crew that inadvertently launched a hellfire missile into the Air Force's ammo storage site?" The brigade commander stood up and said, "I think that's enough, Mr. Bailey. We will be using your aircraft parking plan. Any questions?" There were none.

RETIREMENT

After Desert Shield Desert Storm, I served a bit longer, then submitted my retirement papers for April 1, 1995. I thought April Fool's Day was a perfect day to conclude my military service. That way it would coincide with a special day on the calendar and remind me of my luck in surviving and remember those who did not. I retired on April 1, 1995.

My father was in WWI in the Navy. He signed up for the duration of the war, he went to war, and then the war was over. So he got off his ship and went home. You know, he didn't know there was some kind of out process to leave. The military's like that. I didn't know you had to out process either.

They tell you how to brush your teeth but they never said that at the end, there's a period of time that you have to go through and sign this and get this. I was there for thirty years, but when I got to the end of my thirty years, I said, okay, I quit.

At the time everyone was concerned that soldiers in the Army were getting a bad deal. You know, with Iraq, they drew all these guys in and a year or two after Iraq, now they were getting rid of some of the guys. So Congress passed these laws that the Army had to provide counseling teachers to do interviews. I was in a class with E-3, E-4, E-5, E-6, E-7, E-8 and four colonels.

All these people were at the same level in the class. With this woman and this guy out there teaching us the proper way to interview and how to properly dress for an interview. I remember I looked over at a full colonel, and we just shared a look. It was amazing. But I can understand some of these young guys have been in the Army two or three years and are getting kicked out when they came in right out of high school. They probably never interviewed. But in the Army, if they make a rule, everybody's going to do it and everyone's going to do it right.

After the interview class then you had to go to the Veterans Administration Hospital to see a VA-certified or some kind of doctor and get a full physical exam. I walked in the place thinking I was in pretty good shape. I walked out disabled. I get disability as part of my retirement package. They didn't tell me what's wrong with me but I got this paper in the mail from the VA that said, Congratulations, you have been awarded a twenty percent disability. I think it's because of my back.

Overview of Awards & Decorations

There are things I tried to write, but deliberately did not include. They are difficult to talk about for me and proved harder to write about. They are the things that make nightmares come to life.

You asked for my awards and decorations, and positions or jobs held during my military career.

Awards and Decorations: Legion of Merit, Distinguished Flying Cross, Soldiers Medal, three or four Bronze Stars, Air Medals, no idea how many, one with-V device, two or three Meritorious Service Medals, two or so Army Commendation Medals, a National Defense ribbon. In all, I have been awarded about twenty ribbons. Three unit citations for combat operations. I really don't know all of them. Just read my DD 214. Awarded Combat infantryman badge 1967/68. Awarded Parachutist Wings in 1965, Aviator Wings in 1971 or thereabouts and Master Aviator Wings in the 1980s.

Positions I was assigned in the Army: This is probably not all-inclusive. I have been an infantryman, infantry fire team leader, an infantry squad leader, infantry scout, hand-to-hand combat instructor, jungle survival instructor, improvised explosives instructor, an NBC NCO (Nuclear, Biological and Chemical), a paratrooper, a jumpmaster, a supply sergeant, an operations NCO and warrant officer, a wheeled vehicle operator, M 60 Tank Commander, a helicopter pilot flying utility, scout and attack helicopters, instructor pilot, instrument flight examiner, standardization instructor pilot, aviation maintenance officer, and an aviation safety officer.

I graduated High School in 1965 at age seventeen. I graduated Magna Cum Laude from Embry Riddle Aeronautical University in 1993, after taking college classes for twenty-five years.

The position I am most proud of was being an infantryman. All other forces of the Army, Air Force and Navy are there to support the infantry. It is the infantryman that must close with and destroy the enemy's ability to resist. It is they who see the enemy eye to eye. They must kill, wound or capture the enemy, or be killed, wounded or captured themselves. No other branch sees war like the infantrymen do. They are looked on by some in the Army, including me, as the bravest of the brave. Of course, they have no choice. Once you are in the fight it is kill or be killed.

Above all my other jobs and or titles, I am most honored and proud to have served with the 1st Battalion of the 503rd infantry, in the 173rd

Airborne Brigade, separate from November 1967 to November 1968. The rest, I think you know. If not, just ask.

Our lives are what we make them. In my case it was exactly what I wanted it to be. Full of learning, adventure, challenge, and very little boredom.

PART III: YOU & YOUR MOTHER

I met Donna in Kitzingen a couple of years prior to my divorce when she came to Germany to work for Prudential. I was the standardization instructor pilot in my unit. We were doing some pretty hairy training at the time. Teaching no-light autorotation training at night was dangerous and it took at least twelve deaths for the aviation branch to cancel the requirement.

I would cut the throttle off, like idling your car engine, then we would start a rapid fall toward the ground. The idea is to keep the rotor blades turning so you can soften the impact as you hit the ground. Not crash but skid to the ground, keeping both the helicopter intact and all on board alive. It is risky and I had to allow a pilot to try it at night with no lights and be ready to take control if he was doing something wrong. Which happened often.

I had a few really close calls and figured I owed it to Melodye to increase my life insurance.

On a bad weather day, we standardization instructors were having breakfast in the cafeteria and in walked this guy and a young woman. They walked up to us and he said good morning like he knew us, and introduced Donna as the new Prudential life insurance agent. After they left to get in line for food I asked if any of the guys knew him. No one did.

That night I had another hard landing. I thought God was telling me something. The next morning, I was having coffee in the cafeteria when I saw Donna sitting at a table. I walked over and introduced myself and told her I had a policy with Prudential but wanted to increase it by a hundred thousand. I gave her my number and asked her to call my wife and set up an appointment.

I forgot to tell Melodye. She asked who Donna was, and I said, "I give up, who is Donna?" She got angry and we fought. A week or so later she asked for a divorce, but it didn't really have anything to do with that. I never cheated on her, but my job did cause a lot of protracted separations and she eventually fell in love with her boss. It was amicable and final in about three months.

I ran into Donna off and on around the base or at military functions. When we saw each other, we would talk about her ski trips and make small talk. I dated a few women. The last one was a Japanese lady named Maureen. About a year or so after my divorce, Maureen and I were in Heidelberg for the aviation ball and we ran into Donna and her date, who was the battalion flight surgeon.

We talked for a few minutes and when she left, Maureen went a little nuts. Who is she and how do you know her, etc. She said, "That woman wants you." I laughed and told her she was wrong. "I have known her for a while and there is nothing between us. Besides, she is way too young for me."

Then Donna asked me to hold her purse and camera while she danced and Maureen went ballistic. Our relationship soured. Maureen was a little high strung for me. It was bound to happen sometime.

Then I met a young lieutenant at the House of Lights, which is a country and western place. She was six feet tall and attractive. She could really dance. We dated for a short time. She was a signal officer. She asked me once if I thought she could go to flight school. I said, "Let's find out. I will pick you up tomorrow at the helipad by your site and we'll see what you've got."

The next day, I flew out and landed on the pad and asked for her. She came out and said, "Are you crazy?" I said, "Are we going to do this or not?" She said, "Give me a minute." I guess she had to tell them something. She got in and I took her up for about an hour. She did okay. I told her, "If you're serious, now would be the time to apply for flight school. It takes a while to go through the system."

We stopped seeing each other when I met a female sergeant who was very attractive and more my size. We dated until I left for the senior course, which is six months of classes and supposedly very selective.

Later I ran into the lieutenant, who was now a major and a Chinook pilot, in Saudi Arabia. She was in the 101st Division aviation brigade, to which we were attached. She had gotten married and divorced. I told her about Donna and you.

After the divorce from Melodye, I got a letter from the first life insurance company I had bought a policy from when I was in basic training for my mother. They had been bought and they wanted me to change my allotment to the new company. I was now single and really had no contact with my family.

So I called the flight surgeon and asked for Donna's phone number. He laughed and gave it to me. I wondered what he found funny. I called and left a message explaining the situation and said that I was not interested in insurance but wanted her advice on what to do. I offered to make her dinner in exchange for her time.

She called back a few days later and came over. She read the policy and explained my options. She said it really was not a good policy and she recommended canceling it. We ate and she started flirting with me. This kind of caught me by surprise and I said, "Donna when I met you I was married, I never thought of you as a woman before I just thought of you as the insurance lady." I thought she was a little young for me.

We started dating anyway. On the second date she told me we would date for a while and when it was over, she did not want me to call or bother her. I smiled and asked if I could give her some advice. She said sure. I told her that I was due to leave in a few months and her request was fine with me. "But I think you should wait a little longer than the second date to give this speech again because someday you might fall in love and want to get married."

Ella Fitzgerald Concert in London

We were having a great time and things became serious. We talked about marriage and even looked into having it in Germany. But I broke out in a cold sweat at the Polisina Hotel in Frickenhausen when we were looking at venues and Donna said let's slow down.

She'd read about the Ella Fitzgerald concert in London and said she would love to see it. I bought tickets for the concert about a month away. I showed her the tickets and said, "Let's drop this marriage talk, and if I decide to ask you, I will do it there." She said that was a good idea.

She asked if I had a suit. I told her I only had dress greens and blues. She took me shopping and I bought my first suit with all the trimmings.

We flew to London. The night of the concert we took a taxi to the hotel it was being held at. We'd asked the taxi driver to take us to the hotel where the Ella Fitzgerald concert was, and he said yes and dropped us off. There was a huge crowd. We got out and walked up the red carpet. There were TV cameras, flashbulbs and cheering. Then suddenly everything got quiet. We walked through the door and were escorted into a small room with a greeting line with four or five older women greeting their guests.

We passed the line and as we did, I heard the hostess say, "Oh yes, my dear, I knew your mother." As we walked by, all the ladies in the line looked at us, puzzled. I got two glasses of Champagne and we stood in the rear. I told Donna I did not think the concert would be in this small room and I would be right back.

I walked out to one of the men who had shown us in and asked if we were in the right place for the concert. When he turned, I noticed he had an Uzi under his coat. He said, "Aren't you with Lady so and so's party?" When I said no, he grabbed my arm and I knocked it away, saying, "My date is in there." I went back to get Donna and he escorted us through the bowels of the hotel to the concert.

We'd avoided the very long line and were escorted to a table for eight or ten. We were early and only a few others were seated. The room was huge and the tables were set up in a C shape with the stage maybe six feet above the floor. We were seated right in front—prime seats. The first people to arrive at our table was a couple and their two young girls.

It took some time to fill the room. Once our table was seated, a man came over and asked if we would like our wine served. I had pre-ordered some wine and told him yes, as did the others. He came back with one of ours and told the others theirs had not yet arrived. I asked him to bring all of mine and invited the rest of the table to please join us.

This broke the ice; they were a delightful group. As we introduced ourselves, Donna told them she was originally from New Jersey. They all were from England. The man with the two girls said he had been to New Jersey and thought it a beautiful place. It was then that a teenage girl tapped Donna on the arm and asked her to ask him if he was so and so Pete Townsend, from The Who. She did and the girl smiled and returned to her table, where everyone looked at the man and smiled. Donna said, "I remember when you were in New Jersey. I saw you at The Capitol Theater. It's is nice to meet you." He thanked her and said likewise.

We were among the rich and famous. When I asked how they had gotten into the hotel, this seemed to puzzle them. So I told them about our entry and that we had no idea of how to get out, since the way we came would be impossible to retrace. The table went quiet. They were all staring at us.

Then Pete Townsend said, "You actually walked down the red carpet and went to the lady so and so's reception?" I said, "I do not know whose reception it was, but yes, with a bunch of cameras and mobs of people. The security escorted us here with a man carrying an Uzi." Another moment of blank stares. Then they all started laughing.

Seems the red carpet in England is only for royals and this was a gala charity affair hosted by Princess Margaret, the sister of the Queen. That's why the press was there.

The dinner and the concert were fantastic. We took a taxi home and when we got inside, I got down on one knee and asked her to marry me.

Marriage

We were married on December 29, 1984. Meeting your aunts, uncles and grandparents on your mother's side was my first indication of how vast the differences were between our life experiences and philosophy of the government's role in our daily lives. I did not feel all that welcome or comfortable. (Still don't.)

An example of philosophical differences came about when your Aunt Dianne, your mother and I went for a jog in a park near her apartment. My life has been one of readiness to confront potential hazards and to eliminate them or at a minimum neutralize them. That day, as we ran, the two of them were deep in conversation. I noticed two young men running toward us and showing no signs of letting us pass. I got ahead of them and sure enough the two guys kept ignoring our approach. I was about ten feet in front when we collided and I knocked the closest one down. When he got up, I realized their intent was not hostile, they had just been absorbed in conversation. I relaxed my stance and apologized. Dianne was sure I had been deliberate in my action, but I denied it. (Note: I have never fessed up. So do not share this with her until I am buried).

The wedding was both beautiful and frightening. My mind was on not screwing up and thinking that I had only been single for a short time and suddenly I was responsible for another person's welfare and happiness. As you know, I take my commitments seriously.

We flew back to Tacoma and into our new life. It was a little bumpy. She was not happy with the floors and we needed some new furniture. We did a lot of shopping, which is not my favorite thing to do. We went through a lot of money and when I was balancing my checkbook, I casually said that we needed to slow down, there was more going out than coming in. She started crying and I realized I had hit a sore point. Her father had controlled the money and gave her mother an allowance even though she had her own job. One of your mother's fears is that she would marry a control freak like her Dad.

I thought about it and made her a deal. "I will give you the checkbook, make it a joint account and that way you will know what we have. You run the finances and stick with me and we will have a great time and when I retire you can pick where you want to go and get the house you want." And that is exactly what we did.

I never met Donna's father. But he sounds like a real controlling man. I think your aunts had the same fear. She was happy handling the finances and it really made no difference to me. Between us, she needs the control to feel secure. I have few needs and have relied on myself since I was a kid. I want her to be happy and to be and feel safe.

In hindsight, I should have known she had a problem with money. She wanted to get a prenup and insisted she would pay half of the joint expenses. She must have trusted me, because I asked after a year of marriage if she still wanted one. We could get a postnup and split everything fifty-fifty, but I do not remember her ever bringing it up again.

I know you think your mother is hard, but really she is very soft. Her life as a kid was in some ways much harder on her than mine was. She defends herself by acting tough and uncaring so no one will know if they hurt her.

You will find almost everyone has two sides. The one they portray to the world at large to hide their scars (everyone has scars). It is part of the human condition. My way of coping is to shut myself off from most of the world. I learned and have shared my three circles of philosophy with you, but you may not remember. It goes like this. I have three circles in my life. The inner circle are those things I can control. The next circle is those things I can influence. The last is all things I have concern for but can neither control or influence. Examples:

Control: I can control the relationship with your mother by being the best husband, confidant, friend and supporter I know how to be. I know I spoil her. I tried to spoil you, but somewhere in junior high or high school you really grabbed the controls and became your own person. I admit for me it was a little scary. But there was also some pride. Not in all the things you have done but, in your spirit, determination and intelligence. I hoped that by sharing my thoughts I could influence your attitude and give you the means of dealing with life and its uncaring nature. I wanted you to feel confident about being who you want to be and realizing no one, I mean no one, knows how you feel, your dreams or aspirations. Only you and you alone can decide how to live your life. I think that is the most important thing a person can do. It must be terrible to wake up in your forties or fifties and realize you missed the chance to try to live your dreams.

Influence: These are things that I can try to push in a direction I will be comfortable with. Here I win some and lose some.

Concern: This encompasses politics, religion, bigotry, famine, war and loss of loved ones. That is why I keep people at a distance, because they are not controllable, nor are they trustworthy. I know, you and your mother know I can sometimes seem over-protective. "Where are you going, when will you be back, who are you going with, do you have your phone?" etc. But you are important to me. I still worry about you even though you are a full-grown woman now. I know that if I bug you, it will be an irritant. I also know how much like me you are. I accept that you are bright and will do what you want to do. That is as it should be, whether I like it or not.

In me, fear turns to rage pretty fast. When I care about someone and they are threatened, I can become a little overly aggressive in their defense. So to limit that, I focus on the people I care about. I hope the rest do okay but I have no desire to add anyone new to my life.

I really have no idea how you, or your mother, and your mother's family and friends see me, nor do I care. I would hate to be a source of embarrassment. But I have never been able to be other than who I am. Your mother will run defense for me if someone is pushing me to join the conversation. She takes very good care of me. And after thirty-eight years of marriage, she is still trying to change me. I call that love.

You

Your arrival was a pivotal moment in my life. You were our second attempt to have a child. We had not been married a year when we discussed it. Donna said, "Everyone says you should wait two years to see if the marriage is going to last." I said, "I think it will," she told me, "Me too." So we went for it. She was pregnant in no time. And we were both a little giddy about it.

Donna got pregnant right away but something went wrong in the sixth month. He was stillborn, with some deformities. We were crushed and I was scared. I had served in Nam in an area where Agent Orange had been used and it had caused a lot of problems with children having serious side effects from it. We went to some specialist doctors to have our genes tested to find out if either of us had something wrong with us. We were both pretty shaken by it.

They told us our child was just an anomaly and that we shouldn't expect any more problems. We were a little concerned but we both wanted a child. And so, you were conceived.

The pregnancy went great. We—really your mother—had regular checkups and we went to classes about the process of having a natural childbirth. Your mother wanted to have natural childbirth without any drugs, as you would be getting them too.

The time finally came for your arrival. It went like clockwork. There was one really scary moment for me when your head was first visible. It looked all wrinkly and it scared the shit out of me. You would think someone would have mentioned it, during all those classes, that your head actually compresses coming through the birth canal.

Your mother asked how you looked, and I was not answering. Then your head literally sprung to full size and you were beautiful. They turned your head and, as God is my witness, you looked right at me with big blue eyes and appeared to grin up at me. I kinda lost it for a moment and forgot about telling your mother what was happening.

As soon as you were all the way out, they started trying to make you cry by snapping their fingers on the bottom of your foot. You wrinkled your face up but would not cry. They kept thumping your little foot and you would grimace but not cry. I was getting a little angry and finally asked, "Is there no other way to make her cry?" The nurse said, "I will do it one more time." I said, "If she does not want to cry, I see no reason to force her." The nurse started explaining that it was important. When you found your voice it was loud and constant. The nurses all smiled at me, but I was not smiling. I said, "What does she need?" They said nothing, that you were just angry.

You were still screaming when they took you somewhere for an Apgar test. I have no idea what that is, but they reported with smiles that you scored very high. Donna and you were both tired and they took you off to get some sleep. I went and called the numbers Donna gave me and told people we had a perfect baby girl named Kaytlin Helen Bailey.

I went back to where your mother was recovering. She was wide awake and said she had to pee. I said I'd get a nurse and she said, "No, I just want to pee." She got up and walked to the restroom and on the way back a nurse noticed her and ran to help her. "You are not supposed to get out of bed and he (meaning me) is not supposed to be here." I kissed your mom and said I would be back in the evening.

I came back at around six p.m. Your mother was not happy. She said, "You need to get me and Kaytlin out of here." I talked to the nurse and was told I would have to get the doctor's approval. I went back early the next morning and talked with the doctor, who told me I would have to sign an against doctor's advice release form. I did. He explained that exposure to others was very risky and we could not have anyone in the house for at least two weeks.

I took you and your mother home with the proviso that we follow the doctor's orders. Your mother agreed and I told her I was serious about it and she said no visitors for two weeks. Neither of us had a clue as to what we were doing, but we knew you would let us know when something was wrong.

That first morning you slept for a long time. You and your mother were restricted to the bedroom and the bathroom. I got you into a little basket and Donna in bed and then I checked your nursery to make sure everything was in order. Then I sat in the living room waiting to be called. You were both sleeping so I would wait a few minutes and then check. You two slept for hours. It was nerve racking.

Finally, you cried and I ran in. Donna was reaching down to get you and I said, "I got this." I checked and you had a wet diaper. I had gone to classes and was confident I could handle this. I took you to the nursery. I went over the procedures I had learned. Everything was laid out on your changing table. I laid you there too and proceeded to change your diaper.

They had warned us about little boys, that you need to cover their penis when you remove the diaper as the colder air may cause them to pee. I smiled to myself, thinking that would not be a problem. I cleaned you up, put on the powder and rubbed your umbilical cord with alcohol. I lifted your legs up to put the diaper under you, feeling very confident, and you pooped.

Not a normal poo but a yellowish, watery poo. Because I had your legs held up, it shot all over my white sweatshirt and arm. I was momentarily dazed. Here I was, with your little hiney in the air and poo all over me and you. No one had told me about this possibility. You seemed very pleased with yourself. It was a little awkward holding you out of the mess and cleaning it up. But I was successful and you were perfectly relaxed the whole time.

Donna called out, "Do you need help?" as it took me a while to get everything squared away. I brought you back and she inspected my work. She tried to be casual about it, but I could see she was not sure I would do it correctly. I had, and you passed inspection. She fed and burped you and you went back to sleep.

When I told her what had happened, she laughed and said we are parents.

Your first few nights did not go well. You slept in your basket on your mother's side of the bed and when you cried, I would get up and check your diaper and if it was good give you to your mother. Donna would feed and burp you and we would wrap you in your blanket and lay you back in your basket. That would only last a few minutes and you would start crying again.

This went on for a couple of days and we were getting worried. Donna called the birthing instructor and we found out that we were not wrapping the blankets tight enough. Seems being cramped in the womb restricted your movement and since you now had no restrictions you would wake yourself up by jerking your arms and legs. After we figured that out, things were going great.

I think it was day ten when Donna asked me to go get something from the store. When I came home, there was a strange lady sitting on the sofa holding you. I was pissed. I walked straight to the kitchen and put away whatever I'd bought. I heard Donna explaining to the woman that she had to go because we were not supposed to have visitors for another four days.

When she left, I came out of the kitchen obviously pissed, and said so. Donna said she was sorry, but the doorbell rang and it was a neighbor friend of hers and she had answered the door with you in her arms. She had only been in the house for ten minutes when I came back. I explained that I loved her and you and I would not be very happy if either of you got sick. So no answering doors, no visitors, no risk for four more days. And that is what your first fourteen days were. You, your mom, our dog Megan, and I.

You had more clothes than either your mother or I did. We got packages from all over the USA and Germany. You went through four or five outfits a day. And you still grew too fast to use them all.

Your mother invited over a few of the couples from our birthing class with their babies. It was great. You and the other babies were about four months old and you all seemed fascinated with each other. I do not remember one of you crying the entire time. You all kept looking at one another. Your mother has a photo of it somewhere. The conversation was a little weird. Both the fathers and mothers talked about giving birth in excruciating detail. I guess that is normal. I felt a little uneasy with it, but they all seemed comfortable.

After that we had constant visitors. Your aunts and uncle from New Jersey, Grandpa Curt and Nana. Your Uncle Dick and Aunt Joanie. And friends. It was like you were a rock star and they all wanted to see, and film, you. I was amazed at your tolerance for all these strange people holding you and how you would look at them as if you were trying to memorize them or something.

The introduction I liked the most was when you were about six or eight weeks old and we finally introduced you to Megan. You and her were something. I walked you into the den and sat on the couch and patted my knee. She was all excited; her whole rear end was wagging. She approached as if she knew you were special. She walked very slowly, head down and nose twitching. She put her head on my lap and sniffed your blanket. I put your hand out and she sniffed it and her tail was going a mile a minute. Then she shocked me. In an instant she brought her head up and licked the side of your face. Then she sat back down, tail still wagging. You couldn't take your eyes off of her. You two locked eyes for a very long time. I thought, "They will get along just fine."

I was at work the day your mother called to tell me you had rolled over. I came home and watched you on your blanket for hours. Nothing. You rolled over again the next day and I caught it. Then you started to get up on hands and knees and rock back and forth. Your mother and I could hardly wait for you to crawl. Neither of us had a clue what chaos this would cause.

I think you were about four or five months old when I came home from work and you had half your arm down Megan's throat. I thought she was biting your arm off. I ran in and separated you. Donna was casually cooking dinner in the kitchen. I said she needed to watch you better because Megan could have bitten her arm off. Donna gave me a puzzled look and asked, "What are you talking about?" I explained what I saw when I opened the door. She said, "They do that all the time. They swap their cookies. Sometimes Megan will steal Kaytlin's and sometimes Kaytlin will steal Megan's." I was absolutely speechless.

It took me a while to gather my thoughts and I finally said, "It has to stop. One of them could get hurt." Your mother looked at me as if I was from Mars and said something like, "Well, dear, you explain it to them."

After that I used to watch you and Megan interact. I would watch when you would start to crawl out of her area and she would push on your hiney and then lay down in front of you, so you would turn and head back into the rec room. Or you'd watch her tail and try to grab it.

Sometimes you got the best of her, and she would come to your mother or me and put her head on our knees and whine. This was her way of saying, "The kid is driving me crazy, please do something." We would let Megan out or put you in your playpen. You and Megan were very close for your first three or four years.

Ft. Hood, Texas

I think you were around four or five months old when I was assigned to Fort Hood, Texas. We packed up the Jeep and put you in your car seat and Megan in the rear and off we went. I would glance from time to time. Megan would have her head lying next to you and if you were not sleeping you would be touching her. It seemed you brought comfort to each other. On that long and often boring trip you never got cranky. You and Megan often shared your treats and I admit not being all that comfortable with it, but you two seemed fine with it.

It was in Texas when you and I had our first battle of wills. Your mother had family there and we stopped to visit them. We went to a restaurant for lunch. Three adults and three children, you being the youngest. You started throwing a tantrum. I think it was because the other two kids were sitting in chairs and you were in a high chair. I picked you up and we went outside. I explained that it was not ladylike to make such a racket in public and we would return when you could behave.

You were crying like I was beating you and screaming as loud as you could. There was not a tear in your eye. You were just angry and wanted your way. When you stopped, I asked, "Do you want to stay out here or go back in?" It was July or August and it was really hot. I explained we'd go back in but if you started acting up, we'd come right back out here. We went in and I put you in the high chair. When I sat down you started to scream. I got up and you stopped. I said, "Next time you do that we will not come back in." You gave me a look I will never forget. You locked eyes with me and I could tell you were thinking. It was only a moment but it told me you were very much your father and mother's child and would always challenge the limits and rebel against authority.

We left and went to a small roadside amusement park. We were the only customers. We went from ride to ride and you picked what you wanted to ride. The man was very nice and we spent a couple of hours with you picking your rides. When it was time to go, I said we have to leave now. I was prepared for another tantrum. But I picked you up and we left. We had driven maybe a mile when I glanced back. You and Megan were both sleeping.

Texas was the first time your mother and I experienced your temper on full display. It was too hot to go outside during the day. We would take you out and play in the yard or take you on a bike ride. You would soak in everything you could. Your head was like a swivel, looking left and then right. In constant motion unless something really got your attention. Then you would try to keep looking at whatever it was for as long as you could.

We passed by this kid's playground on our trips but usually it had a bunch of kids playing. One evening it was empty. We stopped and I would take you up the slide and your mom would catch you. On the swings, your mother would hold you in her lap and I would push the two of you. You were completely quiet and just taking it all in, no smile or giggles, just looking around.

Then we tried to leave and all hell broke loose. I mean your screams were chilling. I stopped to check and see if something had bitten you, but your mother said calmly, "She does not want to leave yet. Head back and she will stop." I turned the bike around and started toward the park and you stopped. I stopped and said, "How did you know?" Donna said, "I am with her more than you and she is our child. Think about it, are either of us willing to not get what we want without a fight? I know this is news to you sweetheart, but your daughter seems to have inherited your and my strong desire to do and have what we want when we want it. And your temper. Which we both know is quick and hot."

I said something stupid and turned the bike around and started home. You screamed until you were too tired to continue, with not one tear shed. I said to Donna when we got home, you know this could be a problem. Your mother smiled at me and said something like, not if we teach her manners and to try and focus her on reaching her goals. I was never really sure from that day until you were about four if some of the things you did were because I or your mother had pissed you off.

The very next day I was putting a grill together on the patio and you wanted to be out there with me, so Donna let you out. I was lying on my back screwing the frame together and not focused on you. Out of the blue you had picked up a large crescent wrench and hit me in the head and immediately took off for the door. Your mother said, "She was trying to help you." I said, "I am not so sure. She is my daughter and I remember my childhood."

My suspicion was confirmed years later in Bad Schwalbach Germany. I came home and was very tired. It had been a tough day. You wanted to show me the pictures you had drawn. I said, "Not now Kaytlin, I am very tired." When I came down you had torn all the pictures up and ignored me all evening. When I got up the next day to go to work, my wallet was not on the table where I left it. I looked for a few minutes and it dawned on me that I would hide my mother's keys if I felt slighted. I woke Donna up around four-thirty a.m. and she was sure I was crazy so she spent twenty or thirty minutes looking before she finally did what I had asked her to do and went upstairs to wake you up and ask you to show her where my wallet was. Shortly you both came down the stairs and I stayed in the kitchen. When I saw you walk up the stairs and go to our room to crawl in bed with your mother, I came out. Your mother was standing with my wallet in her hand and this dazed and confused look on her face. She said, "We have a problem."

I told her I would explain when I got home and I did. It went something like this:

"Our daughter has inherited a strong sense of right and wrong. If she has her way, then it is right. If she does not have her way, then it is wrong. She will hopefully outgrow this to some extent, but if she is like me, she will never totally let it go. I have taken revenge on a perceived wrong done to me many, many, years after it occurred."

Donna looked at me with a kind of deer-in-the-headlights look and said, "This all comes from your side of the family." I agreed and said, "But her looks and charm come from yours." She laughed and we agreed we had a beautiful, intelligent and devious daughter that we love.

I decided then I would have to plant the seeds I wanted to plant and reinforce them as often as I could. Hence the, "You are going to college for four years and graduating," discussions we had from your fourth birthday through your graduation day.

I find it funny that you came to that decision without ever considering all my hard work in trying to sway you. I will take a little credit, just because I find it funny now that I tried so hard for so long. But in truth, if you had not wanted to go, I know you wouldn't have gone.

You have done things I wish you had not done. You have done things I am extremely proud of. And that is how it is supposed to be. When a child grows to adulthood, they become their own person. They make their own decisions and reap the rewards or suffer the consequences of their actions. I would never try to tell you how to live your life. I have no right. I hopefully raised you to be confident in your own ability, to weigh your options honestly, and make your decisions with clarity of purpose.

I hold myself responsible for my actions. No one, since my sixth or seventh birthday, has talked me into or coerced me into doing anything. Nor have I tried to do so to others.

I know I will leave little behind other than you and your mother. Both are far more valuable to me than my life. I think that your mother, and then you, will have the financial resources to live comfortable lives. And that gives me peace of mind.

I have my own set of values and of what is right and what is wrong. My values are rigid, and I do not think I have ever done or been involved with anything I thought was not the right thing to do. This does not necessarily mean legal.

There are probably a thousand things I've left unsaid to both of you. Bottom line is you two gave me the only true joy and an honest feeling of love in my life.

I would appreciate it if you would not use any of this in a way that would in any way make fun of the military. I, of course, am fair game.

With all my love. Your Dad.

End Note

Joseph Edward Bailey died on August 31, 2022. He was seventy-four years old. Joe was buried with full military honors at Arlington National Cemetery on December 8, 2022.